湿地光影
丛书
COLLECTION OF
WETLANDS IN IMAGES

PANORAMIC
WETLANDS

全景湿地

陈建伟
Chen Jianwei
著/摄

中国林业出版社
China Forestry Publishing House

图书在版编目（ＣＩＰ）数据

全景湿地 / 陈建伟著、摄. -- 北京 ：中国林业出
版社，2022.10
（湿地光影丛书）
ISBN 978-7-5219-1889-2

Ⅰ．①全… Ⅱ．①陈… Ⅲ．①沼泽化地－普及读物
Ⅳ．①P931.7-49

中国版本图书馆CIP数据核字(2022)第181858号

出 版 人：成　吉
总 策 划：成　吉　王佳会
策　　划：杨长峰　肖　静
责任编辑：袁丽莉　肖　静
宣传营销：张　东　王思明
特约编辑：田　红
英文翻译：柴晚锁　谭　鑫 文婷婷 肖奕菲 薛逸然
图片编辑：崔　林
装帧设计：崔　林（依丹设计）

出版发行：中国林业出版社（100009#北京市西城区刘海胡同7号）
http://www.forestry.gov.cn/lycb.html
电话：（010）83143577
E-mail：forestryxj@126.com
印刷：北京雅昌艺术印刷有限公司
版次：2022年10月第1版
印次：2022年10月第1次
开本：787mm × 1092mm　1/12
印张：18
字数：160千字
定价：320.00元

目录 CONTENTS

导言 INTRODUCTION ---------- 15

大气水循环
Atmospheric
Water Cycle

1. 大气环流 水循环　Atmospheric Circulation　Water Cycle ---------- 28

2. 热力效应 泛三极　The Thermal Effects　The Triple Poles ---------- 44

3. 亚洲水塔 江河源　The Water Tower of Asia　The Source of Major Rivers ---------- 56

湿地类型
Wetland Types

1. 河流湿地　Riverine Wetlands ---------- 74

2. 湖泊湿地　Lake Wetlands ---------- 84

3. 沼泽湿地　Swamp Wetlands ---------- 94

4. 滨海湿地　Coastal Wetlands ---------- 104

湿地生物
Wetland Life

1. 湿地鸟类　Birds of Wetlands ---------- 112

2. 湿地其他野生动物　Other Wetland Wildlife ---------- 132

3. 湿地野生植物　Wild Flora in Wetlands ---------- 142

生命共同体
Community of Shared
Future for All Lives

1. 湿地与森林　Wetlands and Forests ---------- 156

2. 湿地与草原　Wetlands and Grasslands ---------- 168

3. 湿地与荒漠　Wetlands and Deserts ---------- 176

4. 湿地与海洋　Wetlands and Seas ---------- 186

5. 湿地与人类　Wetlands and People ---------- 194

生态思考 ECO-THOUGHTS ---------- 209

后记 EPILOGUE ---------- 214

大湿地大希望 • 黑龙江嫩江
Vast wetland, grand hope • Nenjiang River, Heilongjiang

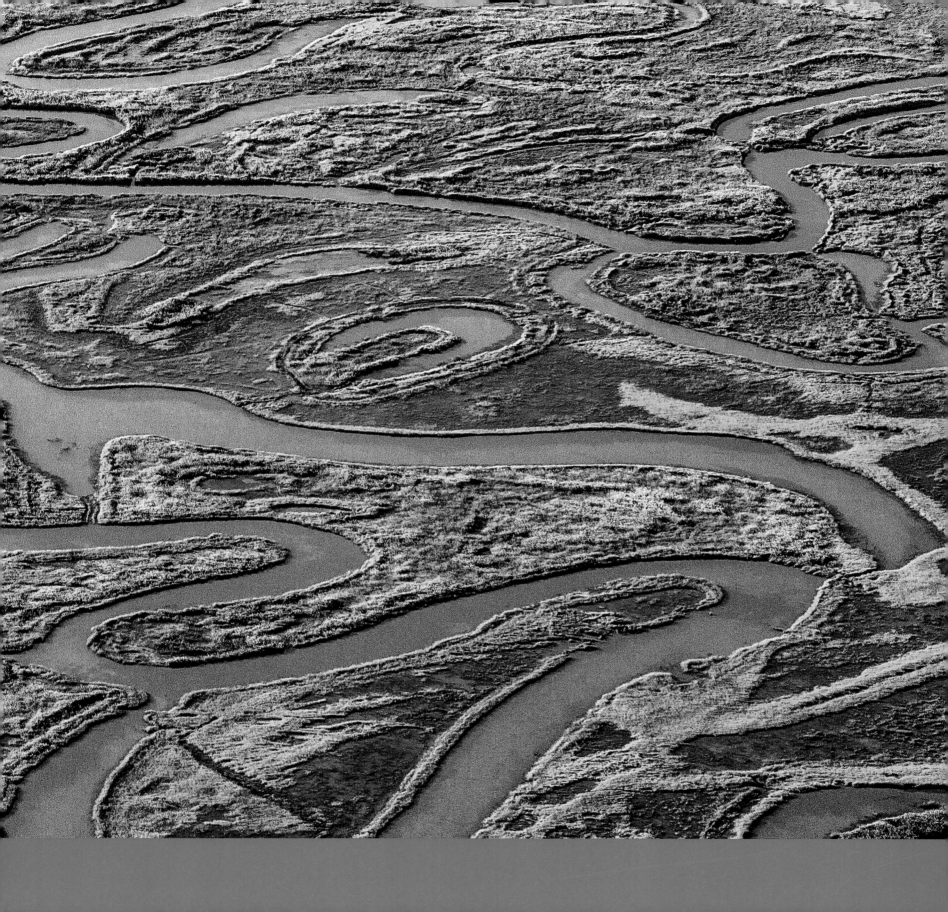

大九湖美丽景观·湖北神农架
A stunning view of the Dajiuhu Lake • Shennongjia, Hubei

世界最大的黑颈鹤繁殖地·四川若尔盖
The world's largest breeding site for the black-necked crane (*Grus nigricollis*) • Ruo'ergai, Sichuan

西溪湿地•浙江杭州
The Xixi Wetland • Hangzhou, Zhejiang

导言
INTRODUCTION

一张图片的启示
An Enlightening Photo

《湿地公约》发展的新趋势
New Development Trends of the *Convention on Wetlands*

水与湿地
Water and Wetlands

篇章结构和图片说明
Notes on the Organization of the Book and Photos

生态系统之大成・新疆哈密
A Masterpiece of Ecosystems • Hami, Xinjiang

一张图片的启示 -

《生态系统之大成·新疆哈密》这张图片向我们展示了一个让人震惊的景观，蓝天、雪山、森林、荒漠、草原、湿地和野生动植物等诸多元素都集聚在了同一个画面中，这种景观不仅在中国难寻，在全球也是极其罕见的。当我拍下这张照片的时候，心里产生了强烈的震撼，同时也引发了我深深的思考。

这是除了海洋（几乎不可能挤进这样的景观里）之外，地球生物圈所有的生态系统要素都囊括进来的景观，它除了引起我们强烈的震撼外，我们有没有想过：是什么内在的元素把它们有机地联系在了一起？

是水！雪山融水可以滋润森林，再往下滋润草原、荒漠，最后都归进了湿地，湿地中水鸟在芦苇丛中游荡。各类生态系统离不开水，野生动植物没有水就不可能存活。

由于重力作用，陆地上所有湿地里的水最后都要归入大海，而沿途被太阳照射还会发生水分蒸发（水分蒸发也发生在所有森林、草原、荒漠中），水分最后都进入了大气层，当然，占地球表面积71%的海洋是水分蒸发的主力。这些进入大气层的水汽会凝结成水滴、雪花或冰晶，藏在云雾中，随大气环流走遍全球，并以降雨、降雪的方式回归大地。这样，就形成了一个闭环、一个完美的大气水循环。地球所有生态系统都在这个水的循环圈中，无一例外。

联合国《生物多样性公约》第十五次缔约方大会（COP15，2021年，中国昆明）的主题是"生态文明：共建地球生命共同体"，《湿地公约》第十四次缔约方大会（COP14，2022年，中国武汉）的主题是"珍爱湿地，人与自然和谐共生"，这些中国理念其实已经成为了国际社会的共识和人类为之奋斗的目标。在我们的自然世界里，地球生命共同体最基本的纽带是水，人与自然和谐共生最基本的纽带仍然是水！在地球所有陆地生态系统中，湿地是水最密切、最重要、最不可或缺的载体。水是生命之源，我们应该更加珍爱水、珍爱湿地，才能创造人与自然和谐共生的新局面。

《湿地公约》发展的新趋势 -

《关于特别是作为水禽栖息地的国际重要湿地公约》（简称《湿地公约》），于1971年2月2日在伊朗小城拉姆萨尔签订，1975年正式生效。《湿地公约》缔结至今已经51年了，随着整个国际形势的发展，环境保护成为当今世界的主流，《湿地公约》关注的方向和内涵也随着时代的发展产生了很大的变化。

回顾《湿地公约》的发展历程，我们可以看到这样的趋势：《湿地公约》最初专注的是水禽栖息地和迁徙鸟的保护，后又增加了鱼类作为衡量国际重要湿地的标准，接着又决定加强对泥炭地、珊瑚、海岸带、喀斯特地貌等类型的保护。之后，更发展到关注水文变化、水质净化、水源提供、水资源管理、流域综合管理等方面。再后，水与人类、湿地生态系统的整体保护、全球气候变暖、湿地文化等，都逐渐成为缔约方大会的热门话题和政策重点。从对作为水禽栖息地的国际重要湿地的保护扩展到了对全球湿地生态系统的保护，《湿地公约》还朝着关注人与湿地关系（包括建设"国际湿地城市"）等方向发展，就连徽标也进行了修改，由飞行的水禽与湿地栖息地的组合到反映湿地的多种功能，它强调的是整体的湿地生态系统，更突出了与水的关系。

我们尤其注意到："水"已经成为《湿地公约》所关注的重点和发展新趋势，以至于有人说《湿地公约》的未来发展应该是"水公约"。"水是湿地的灵魂"，我们确实应该把水和湿地的关系提高到这样的高度来认识。

水与湿地 -

湿地呈现给我们的不只是河流、湖泊、沼泽、滩涂等不同的形态，不仅是因为同一流域的很多湿地之间关系紧密相互影响，或正相关或负相关，更是因为地球的大气水循环，无论湿地形态如何、位置怎样，水都将全球的所有湿地有机地联系在一起，形成了一个统一的、完整的环球大生态系统。

地球上的森林、草原、荒漠、湿地生态系统，在生态景观学上都是斑块状的存在，斑块之间都存在一定的联系，而湿地斑块之间的联系要比其他生态系统斑块之间的联系紧密得多。例如，这块森林消失了，对于另外一块森林的影响不太大，草原、荒漠也大致如此。但是，河流的上游枯竭了，下游也就快完了；河流没有了，通江湖泊、沼泽也就行将消失；山这边湿地多了，那边湿地可能就少了，等等。

除物质流和能量流之外，一个个湿地斑块还通过水鸟的全世界迁徙，紧密地联系在了一起。全球有若干条鸟类的迁徙路线，这些路线跨越了几千千米甚至上万千米。水鸟在迁徙途中休息、觅食的地方，就是分布于世界各地的湿地，迁徙路线就像一根根线，把这些湿地如珍珠一般串联了起来。我们保护某一块湿地，就是在保护整条迁徙路线，某块湿地一旦消失，迁徙的鸟无法及时落脚补充营养，也许就会影响整个迁徙行为，甚至导致整条迁徙路线中断。一荣俱荣，一损俱损。湿地最生动、最完美地诠释了"地球生命共同体"的理念。

除了休戚与共、相互关系密切外，湿地还有一个特点，就是可以上天入地。森林、草原、荒漠、海洋都有它们自己特定的生态区位。例如，海洋只能在海平面以下；荒漠只发生在特定的自然地理气候区；降水太多或太少的地方都没有草原；森林不仅受光、热、土条件的限制较多，而且不同类型的森林的自然分布都有其上限和下限。唯有湿地与众不同，冰雪世界之外的任何海拔高度、任何自然地理条件都有它的存在。

因此，我们可以说，对于湿地生态系统的特殊性、重要性，我们的认识还远远不够。尤其是从生命共同体的角度来看，湿地生态系统和森林、草原、荒漠、海洋生态系统相比，确实是极为特殊的。

这一切其实都是因为水。因此，本书打破了湿地书籍编写的惯例，不是就湿地而谈湿地，也不是一谈湿地就把湿地类型、湿地区划一一罗列。为了体现当今湿地保护新的发展趋势，本书以水开篇，站在更宏观的角度讲述，让读者对全球范围内水的存在、表征、循环有一个全面的认识。水讲清楚了，湿地的特殊性就容易理解了。

2014年，我出版了《多样性的中国湿地》一书，书中提出了自己的中国湿地区划方案，刘兴土院士评价说，这是"首次从生态系统的角度来对中国湿地进行的区划"；同时，该书还对中国湿地类型和湿地野生动植物做了图文并茂的展示，并简要介绍了中国湿地保护20年（1994年至2014年）取得的成就，也对湿地保护提出了自己的生态思考。对于再版，我本人不太愿意，因为时间已经过去8年了，虽然湿地主体未变、主要基础未变，但是如果不能把这些年中国湿地保护的理念提高和《湿地公约》新的变化趋势纳入其中，再版的意义就不大了。

那么，要重打锣鼓急开张？要重新构思一本新的湿地主题的书确实很难，不仅是因为时间紧迫，更重要的是如何突破原有框架做出新意？新书立意如何符合当今中国及世界环境保护发展的趋势？用新理念创作，既要有清晰、深刻的思想内容，还要有自己拍摄的与之匹配的照片，为迎接《湿地公约》缔约方大会在中国首次召开呈现一个全新的作品，确实不太容易。

在绞尽脑汁重新构思本书时，我试图把水的循环及"地球生命共同体""人与自然和谐共生"的立意作为新书突破口，用更宏观、更科学、更创新的思路来阐述湿地。这就得从大气水循环讲起，而讲大气水循环，就离不开地球"三极"。南北极对全球大气水循环的影响我们已经有了共识，而青藏高原对于地球大气水循环的影响我们认识得并不够，可喜的是它正越来越被人们所认识。

大气水循环给地球带来了万物生长、欣欣向荣的无穷魅力，人类也在其中得到了充分的眷顾。我们怎能不珍惜这样运转的地球呢？明白了这些，我们才会由衷地去热爱，才知道如何去保护，创作这本书才有了新的动力和意义。

篇章结构和图片说明 -

有了以上认识，本书的篇章结构就清晰了。

本书共五章：第一章大气水循环，主要从全球的宏观视野来讲水资源，突出青藏高原作用和"泛三极"效应；第二章湿地类型，主要介绍湿地的各种类型；第三章湿地生物，主要讲湿地野生动植物；第四章生命共同体，讲湿地生态系统与其他生态系统之间的辩证关系，包括与人类的关系；本书的最后，专门安排了"生态思考"一章，主要讲我国湿地保护和修复所取得的主要成绩、面临的挑战和今后的展望。需要说明的是，本书中的湿地类型的划分是按照《土地利用现状分类》（GB/T21010-2017）和《湿地公约》分类方法。

每章的开篇都有一段说明文字，具体内容再以生态摄影作品的形式陆续展开，其中，第一章涉及一些新知识，文字量较多。需要强调的是，生态摄影提倡除了用图片告诉读者尽可能多的影像信息外，还有一个画龙点睛的题目，并附上一段简要的说明文字，讲述摄影者的一些认知或生态思考等。尤其是当图片涉及一些科学知识、生物物种时，都会一一注明，以方便读者能够更准确地理解图片所讲述的生态故事。

本书坚持生态摄影以"思想性、科学性、艺术性"为一体的宗旨，用生态文明思想强调生态关系、讲述生态故事是自始至终都应该被坚持和弘扬的。

An Enlightening Photo -

A Masterpiece of Ecosystems • Hami, Xinjiang presents us with a startling landscape that brings into a holistic whole almost all the views of nature, including blue skies, snow-capped mountains, verdant forests, barren deserts, rolling prairies, life-teeming wetlands and a rich variety of wildlife species that inhabit in this vast scope of lands. This sort of stunning view is not only extremely difficult to be found in China, but also makes up an unusual spectacle in the world. In addition to the tremendous emotional impact it had on me when I took this photo, I couldn't help but sink into deep thinking about the wonders of natural ecosystems.

Except for the seas (which is just downright impossible to appear in a landscape like this), this photo contains within its frame almost all the key elements of ecosystems that exist in the biosphere of the planet Earth. Leaving the almost indescribable sensational impacts it may have on us apart, do you ever for a moment stop to ponder over the following question: what inherent element it is that has brought all these marvelous scenes into such an organic whole?

The answer is water! The melted water trickles down from the snow-capped mountains to give moisture consecutively to forests, grasslands, barren deserts before it finally lands on wetlands, which in turn provide shelters for the water fowls that swim care-freely among groves and groves of reeds. There would be no ecosystems to talk about without water, and no life can expect to survive without the nourishment of water.

Due to the influence of gravity, all water existing in terrestrial wetlands is bound to end up into the oceans on the planet, except for the parts that evaporates under the sunshine during the process (evaporation occurs in forests, over grasslands and barren deserts as well) and enters into the atmosphere. Admittedly, it is the ocean that covers over 71% of the Earth's surface that makes up the primary sources of water evaporation. The vapors that enter the atmosphere will condense into droplets, snowflakes or tiny ice crystals in the clouds and travel across the globe along with the atmospheric circulation before being returned to the earth in the form of rains and snows. Thus, a perfect close-loop atmospheric cycle of water comes into being. All ecosystems on the planet fall within the cover of this giant water cycle, there being no exception.

The theme of the 15th Conference of Parties to the United Nations Convention on Biodiversity (UNCBD COP 15, which was convened in Kunming, China in 2021) is "Ecological Civilization: Building a Shared Future for All Life on the earth". And that of the 14th Conference of Parties to the *Convention on Wetlands* (*Convention on Wetlands* COP 14, which is about to take place soon in Wuhan, China in November 2022) is "Wetlands Action for People and Nature". The ecological ideas denoted by the two themes, both of which have their roots deeply embedded within traditional Chinese philosophy, have already become the consensus as well as the lofty aspiration of the international community. Of all the things in the natural world, the essential element that underpins the shared future of all life on the earth is water, the indispensable bond between people and nature is also water. Among all the terrestrial ecosystems on the planet Earth, wetlands make up the most crucial, most intimate and most dependent vessels for the existence of water. Acknowledged as the source of life, water, together with wetlands, deserves to be valued and cherished, which is essential for the harmonious co-existence of people and nature.

New Development Trends of the *Convention on Wetlands* - - - - - -

The *Convention on Wetlands of International Importance Especially as Waterfowl Habitat,* (known more commonly as the *Convention on Wetlands, or* the *Ramsar Convention*), was adopted in the Iranian city of Ramsar on February 2, 1971 and came into force in 1975. Now that the *Convention*

on Wetlands is already in its 51st years, its focal areas and missions are also undergoing notable changes to better fit in with the current global trends of development in which environmental conservation has become the consensus and top priority of the global community.

In retrospection, we will find that the *Convention on Wetlands* has been through the following trajectory: started as an initiative that focused specifically on the conservation of waterfowl habitats and migratory birds protection, the *Convention on Wetlands* later on adopted additional missions that take fish species as standards for determining wetlands of international importance before covering under its mandated mission the protection of peatlands, oases, estuaries, deltas and tidal flats, mangroves and other coastal areas, coral reefs and *etc*. Afterwards, the *Convention on Wetlands* further expanded its focal areas to all lakes and rivers, underground aquifers, swamps and marshes that play critical roles in the study of hydrological changes, water purification, water supply, water resources management as well as the comprehensive management of catchments as holistic wholes. Still further on, the COPs of the *Convention on Wetlands* listed within its policy priorities such topics as the water-people relationship, integrated conservation of wetland ecosystems, global climate change and wetlands cultures. To better reflect this remarkable shift from being an initiative that specialized in the conservation of waterfowl habitats that are of global importance to one that gives top priority to the conservation of all wetland ecosystems on the planet and to optimizing the people-wetland relationships (as showcased by the endeavor to build the International Wetland Cities), the *Convention on Wetlands* modified its logo: from a combination of symbols that features flying waterfowls and their habitats to a new one that highlights the multiple functions of wetlands as integrated ecosystems and the fundamental role of water within these systems.

What's particularly worth noting is that "water" has recently been identified as the new focal area of its mandated missions as well as the new direction of its policy-makings, even to such a degree that someone claims that the *Convention on Wetlands* would possibly be developed in the future into a *Convention on Water*. Somehow exaggeratory though it is, it is indeed reasonable for us to view the relationship between water and wetlands from such a high elevation, given the "critical role" that the former plays in sustaining the health development of the latter.

Water and Wetlands

What wetlands present to us human beings are not merely such variegated geographical forms as rivers, lakes, marshes and tidal flats, or the intimate and reciprocal influences between the numerous wetlands that fall within the same catchment, be such influences positive or negative. More importantly, what makes them especially valuable is the atmospheric water cycle that gives birth to a huge pan-globe ecosystem, in which all wetlands on the Earth, whatever shapes they take and wherever they are located, are linked up through water into a complete and integrated whole.

Viewed from the perspectives of ecological landscape, all terrestrial ecosystems, such as those of forests, grasslands, deserts and wetlands, exist in the form of fragmented patches that are somehow interconnected with each other. Nevertheless, the ties between patches of wetlands are much more intimate and inherent compared with those existing between other types of eco-patches. Take for instance, suppose one stretch of forest is lost, its impacts on other stretches of forests would be relatively mild and not readily visible. It is roughly the same case with grasslands and deserts. But once the upper reach of a river runs dry, gone would be its lower reach. Once a river is lost, all the lakes and marshes to which the river feeds will soon vanish. The expansion of wetlands on this side of a top mountain would often mean the shrinkage of wetlands on the other side of it.

Besides the material and energy flows, each of the wetland patches are also connected with the pan-globe migrations of waterfowls. There are several avian migratory routes around the world, with some spanning for several thousands and even up to tens of thousands kilometers in length. Places where the migratory birds stop over for rest and for forage are often the wetlands that scatter randomly across the world, whereas the migratory routes are like the threads that string the wetlands into necklaces of pearls. The protection of each individual wetland is the protection of the entire migratory route, given that the loss of any single stopover place along the route is very likely to make the birds without access to adequate rest and food, projecting a negative impact on the migratory process, even causing the whole route to fall apart. The thriving or withering of any given part would mean the thriving or withering of the whole. Wetlands present us with a most telling illustration about the meaning of "a global community of all lives on the earth".

In addition to the close bonds that determine the common flourish or failure of each composing element, one more defining feature of the wetlands is its ubiquitous presence in the world, be they high up in the sky or deep down beneath the ground. Forests, grasslands, deserts and oceans all have their respective positions in the global ecosystem, which are often limited to a given geographical zone. Take for instance, oceans come into being only in places below the sea level; deserts exist only in a certain geographical and climate belt; there would hardly be any grasslands in places with too much or too little precipitations; the survival of forests is often subject to the presence of favorite sunlight, heat and soil conditions, not to mention the upper and lower limits that determine the natural distribution of a given type of forests. But wetlands stand out alone, for they can be found at all altitudes and in almost all natural geographical settings except in permanent snow- or ice-clasped world.

For these reasons, it is reasonable for us to say that our understandings about the importance and peculiarities of wetland ecosystems are still far from being adequate. Judged from the perspectives of "a shared community of all lives", wetland ecosystems are indeed very different from other ecosystems composed of forests, grasslands, deserts and oceans.

And the root cause for all these differences is water. That explains why I am, in compiling this eco-album, not content with following established conventions in previous books that either treat wetlands merely as isolated cases or simply present the readers with lengthy lists of wetland types and distributional zones. Instead, to better align to the new trends in wetland conservation efforts, this book starts with an in-depth look into water and takes a more global and holistic approach in its presentation, which hopefully will give the readers a full and complete picture about the existence, manifestation and circulation of water on the planet Earth. Once a full picture about water is in place, it would naturally become much easier for us to gain a better understanding about the peculiar features of wetlands.

In my 2014 eco-album entitled The *Diverse Wetlands of China*, I presented readers with my own scheme for delineating the wetland zones in the country, which, to borrow the remarks of Academician Liu Xingtu, "marked the earliest endeavor to delineate the wetland zones in China from an eco-system-based perspectives." In addition to that, the album also presented a vast number of vivid pictures featuring the different types of wetlands in the country as well as the rich diversity of wildlife species inhabiting there, which were also completed with a brief introduction about the major achievements that China had made over the 2 decades (ranging from 1994 through 2014) in wetland conservation and about my viewpoints concerning wetland conservation. I personally am not very keen on having this album republished because, though the primary backdrop and foundations of wetland conservation have not changed much over the past 8 years, it wouldn't be of much value to have it republished without a significant overhaul in its contents and overall organizations so as to update the readers with the notably enhanced wetland conservation awareness of the Chinese people and with the emerging trends in the *Convention on Wetlands*.

So would it be better to start all over anew and begin from the ground? It is indeed not an easy job to conceive of a brand-new wetland-themed book, not just because of the pressure of time available, but more importantly because of the inherent challenges that writers are often faced with in breaking away from the confinements and limits of their previous works on the same topic. How can the new book be planned and organized to better align with the current trends in environmental protection in both China and the world? It is indeed a challenging task to come up within the limited time a new book that is innovative in conception, ground-breaking in ideas, and is sufficiently backed up with matching photos taken by the writer himself, so as to celebrate the debut convening of the Conference of Parties to the *Convention on Wetlands* in China.

While I was racking my brain to come up with an innovative approach to the organization of the new book, it occurred to me that it might be a good idea to, by drawing on "the shared future community of all life on the earth" (the theme of UNCBD COP 15) and "harmonious co-existence between people and nature" (the theme of the upcoming Convention on Wetlands COP 14), take the water cycle as a starting point in presenting the wetland ecosystem in a more global, science-based and original manner. This would make it necessary for us to look into the nature of the atmospheric water cycle, which would be incomplete without tapping into the "third pole" of the Earth. While the important roles played by the North and South Poles of the planet in regulating the atmospheric water cycle have been widely acknowledged, that of the third pole—the Qinghai-Tibet Plateau—has not yet been fully appreciated, despite the encouraging trend, to our delight, that more and more people are coming to realize its value.

The existence of the atmospheric water cycle gives rise to a flourishing Earth teemed with the charms of rich biodiversity, which is surely also a great blessing to human beings. How can a person not be touched to value and cherish an Earth that runs on its natural course? Only when all people are duly aware of the innate mechanisms of something, could they be expected to love it from the bottom of their hearts and in turn make conscientious efforts for its conservation. This might be taken as the momentum that inspires me to writing this book and where its meaning lies.

Notes on the Organization of the Book and Photos ------

With the help of previous introductions, a clear picture would emerge concerning the overall structure of the present book.

The book is composed of 4 chapters: Chapter 1, Atmospheric Water Cycle, focuses on an introduction about water resources from a global viewpoint, highlighting in particular on the functions of the Qinghai-Tibet Plateau and the so-called "the 3rd Pole Effects"; Chapter 2, Wetland Types, presents readers with a rough survey of the major wetland types; Chapter 3, the Wetland Life, features in particular on wildlife flora and fauna species for whom wetlands make up homes; Chapter 4, Community of Shared Future for All Lives, explores the dialectical relationship between wetland ecosystem and other ecosystems, including its relationship with human beings. A special section entitled the eco-thoughts is arranged at the end of the book to give a comprehensive review about the major achievements and challenges that China has scored or is facing in its wetland conservation and restoration endeavor, which concludes with a few prospective remarks about future trajectory of

development. It should be noted that classification systems proposed by the *Classification of Land Use Status* (GB/T 21010-2017) and the *Convention on Wetlands* are adopted in this book for the delineation of wetland types.

Each chapter in the book starts with one or several short explanatory paragraphs and is followed up by a systematic display of eco-photographs. The explanatory paragraphs in Chapter 1 are relatively long given that it involves comparatively more knowledge that might be new to readers. It is worth noting that the key point of eco-photographing lies in presenting readers with, an eye-catching visual image apart, a punchy caption and a few brief explanatory lines that reflect the photo-taker's ecological thoughts and particular feelings at the moment when the photo was captured. For this reason, clear notes will be given in this book below each picture, particularly when the picture concerns some specific knowledge or a particular species, so that the readers will more readily and accurately understand the ecological story behind the photo.

While strictly observing to the fundamental principle of eco-photographing that stresses on the seamless integration between ideas, science and art, the book consistently highlights as its prioritized task to promote ecological wisdoms through telling thought-provoking ecological stories.

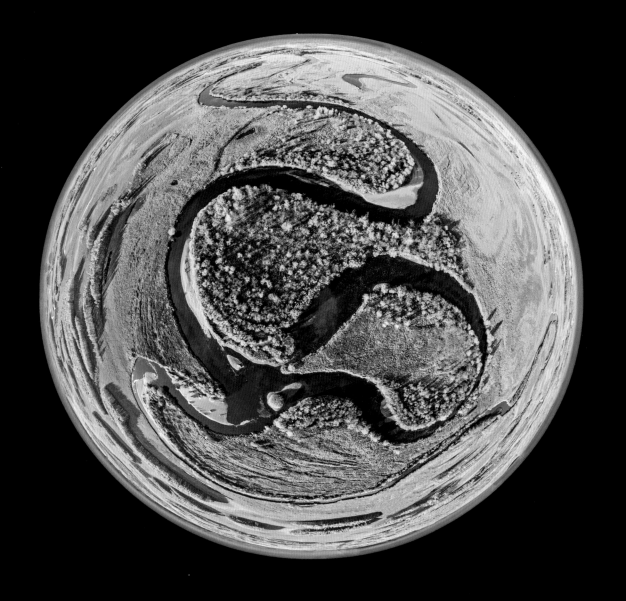

大陆湿地·航拍地球
Continental Wetlands • A Satellite View of the Earth

- -

　　除了海洋之外，占地球表面29%的陆地上，处处分布着河流、湖泊、沼泽和滩涂等人们称之为湿地的地方，人类逐水而居，社会依水而发展，我们的生活离不开湿地。图片显示的是"龙"形湿地。

　　地球任何地方的湿地都离不开太阳的眷顾，在太阳的照射下，湿地中水分蒸发随时都在发生。

　　Apart from the oceans, the terrestrial parts of the planet Earth, which takes up around 29% of its surface, are strewn with rivers, lakes, marshes and tidal flats that we commonly refer to as the wetlands. As proven by that fact that human settlements often spring up in places with abundant water and lush grasses, and that civilizations typically come into being alongside major rivers and water-bodies, wetlands provide us with indispensable shelters on which our very survival depends. The photo depicts a continental wetland that comes in the shape of a giant dragon in traditional Chinese mythology.

大气水循环

　　水是地球上唯一能将固、液、气三态同时展示的物质。中国古代哲学家老子说："上善若水，水善利万物而不争。"蓝色的地球就是一个水的世界，那么地球上的水是怎样交流循环的呢？我们常说"水为生命之源""水是湿地的灵魂"，要说清楚湿地，就必须先从水讲起，从地球的大气水循环讲起。

　　Water is the only substance on the earth that can exist simultaneously in solid, liquid and gas forms. Lao Tzu, the ancient Chinese philosopher, said, "The highest goodness is like water, which benefits all things and does not compete." The blue Earth is like a world of water, so how does the water on the earth exchange and circulate?We often say that "water is the source of life" and "water is the soul of wetlands". To explain wetlands clearly, we must first talk about water and start with the Earth's atmospheric water cycle.

1 大气环流 水循环
Atmospheric Circulation Water Cycle

占地球表面71%的海洋与陆地上的河流、湖泊、沼泽和滩涂等湿地，在太阳的照射下，有一部分水变成蒸气释放到大气中。由于地球的自转和轴心的倾斜，太阳的光热使陆地、海洋表面产生了温差，形成了大气环流和季风。

水蒸气上升形成了云，云被季风带到了世界各地，因各处大气环流、季风以及地形、纬度的不同影响，云变成了降雨、降雪而回归大陆和海洋，从而完成了整个地球的大气水循环，并周而复始。于是，就有了生机勃勃的地球生物圈，有了生物多样性，才有了我们人类。

我们生活的地球是一个蓝色的星球，水的世界。由于大气层的存在，地球有多大，水的舞台就比地球还要大。

The oceans, which cover 71% of the Earth's surface, and wetlands such as rivers, lakes, marshes and tidal flats on land, are exposed to the sun, and part of the water is turned into steam and released into the atmosphere. Due to the rotation of the Earth and the tilt of its axis, the sun's light and heat cause temperature variations between the land and ocean surfaces, leading in result to the formation of atmospheric circulation and monsoons.

Water vapor rises to form clouds, which are carried around the world by monsoons. Due to the different effects of atmospheric circulation, monsoons, topography and latitude, clouds become rain and snow and return to the continents and oceans, thus completing atmospheric water cycle that covers the entire Earth, and the cycle goes on and on. Thereupon, a vibrant biosphere of the Earth comes into being, on which the biodiversity and we human beings depend.

The Earth we live on is a blue planet—a world of water. Due to the presence of the atmosphere, no matter how big the Earth is, the stage of water is bigger than the Earth.

大海蒸腾 • 坦桑尼亚
Sea Transpiration • Tanzania

这是非洲东海岸的印度洋，赤道灼热的太阳像火炉一样烤着海洋，水汽在大海的表面升腾着。

地球71%的表面被海水覆盖，在太阳的照射下，产生了大量的水汽，水汽蒸腾上升后被地球表面的环流、季风等带到世界各地，从根本上影响了全球各地的气候变化。

This is the Indian Ocean on the east coast of Africa, where the scorching sun at the equator is baking the sea like a furnace, and the water vapor is rising on the surface of the sea.

Seventy-one percent of the Earth's surface is covered by seawater. A large amount of water vapor is generated by the solar irradiation. The vapor rises and is carried around the world by the Earth's surface circulation and monsoon, which fundamentally affects climate change around the world.

水汽升腾·黑龙江图强
Water Vapor Rising • Tuqiang Town, Heilongjiang

初升的太阳照在林中河流上，河面的水汽升腾现象在逆光状况下看起来尤为明显。其实，在阳光照耀下，大地的蒸腾作用随时都在进行着。

在广袤的大陆上，在阳光的作用下不断地进行着水汽蒸发的，不仅是河流、湖泊、沼泽、滩涂等有明水面的湿地，森林、草地、农田、荒漠等没有明水面的地方，也都每时每刻进行着水汽蒸发，只是人们平时不太容易观察到罢了。

When the rising sun shines on the river in the forest, the water vapor rising on the river surface looks especially obvious under the backlight condition. In fact, the transpiration of the Earth is going on all the time under the sunlight.

Over the vast continents, the evaporation of water into vapor is taking place constantly under the sunlight not only in wetlands like rivers, lakes, marshes and tidal flats where water is obviously visible, but also in places where obvious water surface is not apparently visible, such as forests, grasslands, farmlands and deserts where the evaporation is not readily observable to human eyes.

水汽逼近五老峰 • 江西庐山
Wulaofeng Peak Veiled behind Misty
Water Vapor • Lushan Mountain, Jiangxi

　　在上午阳光照射加温后，鄱阳湖的水汽开始不断上升直逼庐山的五老峰。五老峰海拔1436米，高出鄱阳湖湖面1400多米，为庐山最雄伟奇险之处。

　　人类大部分居住在低海拔地区。图中景象是在低海拔地区常见的。

After being exposed to and heated by the sunlight in the morning, the water vapor of Poyang Lake begins to rise straight to the Wulaofeng Peak, the most majestic, strange and dangerous part of Lushan Mountain that stands approximately 1436 meters above sea level, or, over 1400 meters higher than the surface of the Poyang Lake.

Most humans live in low altitudes. The scenery in the picture is a fairly common view in low altitude places.

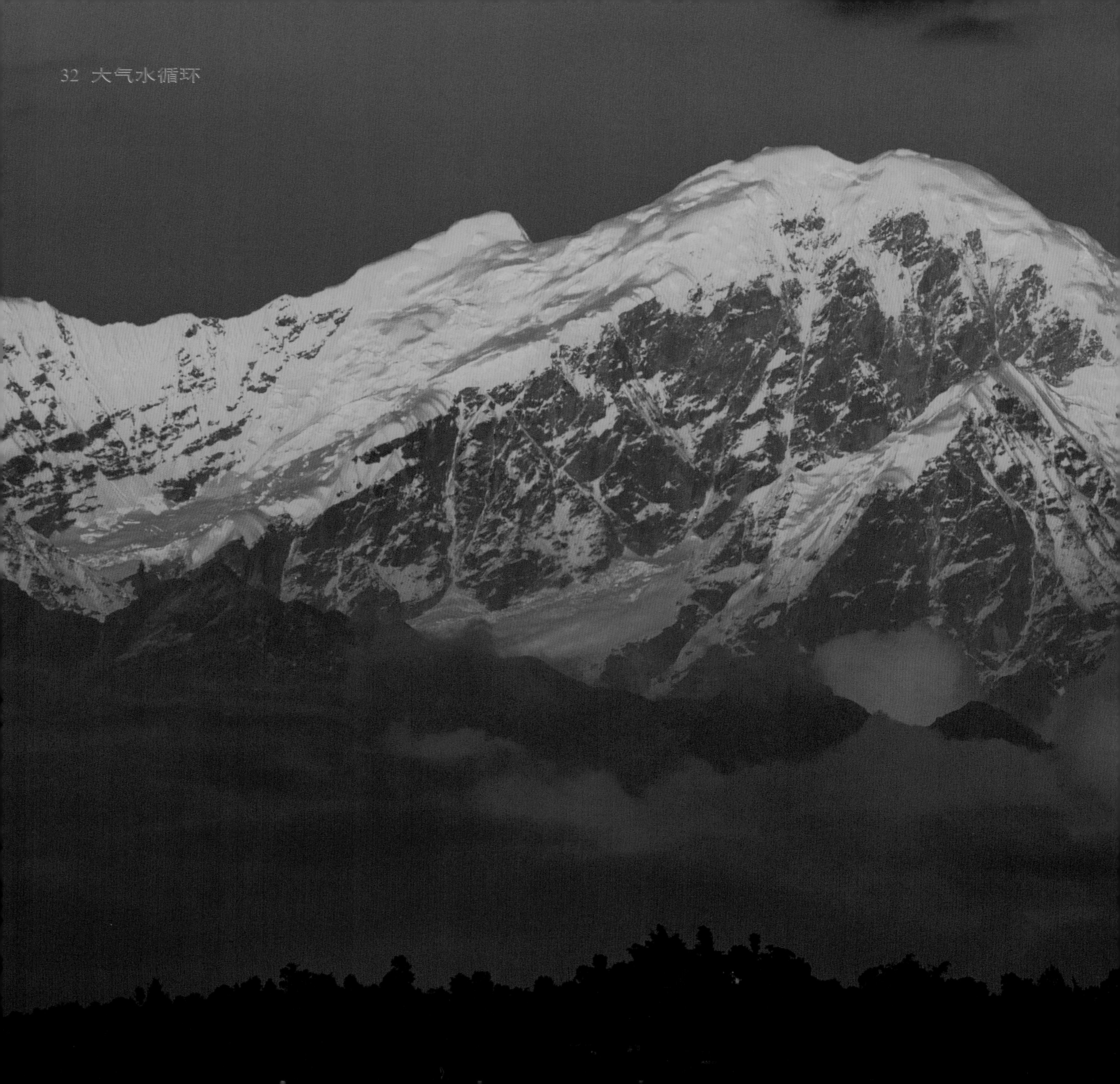

日照金山 • 尼泊尔廓尔喀
The Alpenglow • Gurkha, Nepal

- -

　　印度洋的暖湿气流来到了青藏高原南麓，平均海拔6500~8000米以上的喜马拉雅山脉阻挡了它们的去路，这些云雾只能继续攀爬，并一路以降雨、降雪的水分大量损失为代价。天还未亮，早起的一缕阳光已经出现，把险峻的山体和高山冰川一起刻显、染红。

　　这是喜马拉雅山脉的拉姆加格峰，海拔6966米，已经超过了世界的绝大多数地方。图中景象是在高海拔地区常见的。

The warm and humid airflow from the Indian Ocean arrives at the southern foot of the Qinghai-Tibet Plateau, where its way is blocked by the Himalayas whose average altitude ranges somewhere between 6,500–8,000 meters. These clouds can only continue to climb, at the cost of a large amount of water loss from rainfall and snowfall.Before dawn, a ray of sunshine in the early morning has appeared, and the precipitous mountains and alpine glaciers appear in sharp relief and are tinted in a crimson gleam.

This is the Ramjag Peak in the Himalayas, with an altitude of 6,966 meters, which is already higher than most places in the world. The scenery in the picture is a common view at high altitude places.

大陆冰川 • 斯瓦尔巴
Continental Glacier • Svalbard

青藏高原之外的两极就是我们熟知的北极和南极，这两极基本上被大陆性冰川覆盖。图中是北极圈里斯瓦尔巴岛的内海湾大冰川，从山上一直延伸进海里。

冰川呈现的蓝色和它的成因有关。冰川是由长年累月的积雪压缩形成，积雪中含有很多空气没有逸出，蓝光由于波长较短，在空气中发生散射，因而冰川呈深浅不一的蓝色。

In addition to the Qinghai-Tibet Plateau, the other two poles are known as the North Pole and the South Pole, which are basically covered by continental glaciers. The picture shows the large glacier in the inner bay of the Svalbard Archipelago in the Arctic Circle that stretches all the way from the mountain down into the sea.

The blue color of the glacier has something to do with its formation. The glacier is formed by years of snow compression, which contains a lot of trapped air. Blue light is scattered in the air due to its short wavelength, so the glacier appears in different shades of blue.

路在何方·北极
Where Is the Road Ahead? • The North Pole

　　全球气候变暖，冰川融化，这对北极熊的生存产生了极大的影响。图中的北极熊母子在破碎的冰块上遥望远方，表情茫然，好像在问："今后，路在何方？"

　　所谓的"北极大陆"并没有陆地，北极是北冰洋上漂浮的冰川组成的大冰盖，如果全球气候变暖持续下去，北极的大冰盖不断融化，"北极大陆"将不复存在。

Global warming and melting glaciers have had a huge impact on the survival of polar bears. In this picture, a mother and baby polar bear look into the distance on the broken ice, with a blank expression, as if asking, "Where is the road ahead?".

There is actually no land on the so-called the "Arctic Continent". The Arctic is a large ice sheet composed of floating glaciers in the Arctic Ocean. If global warming continues and the large Arctic ice sheet continues to melt, the "Arctic Continent" will cease to exist.

冰盖碎裂与融化加速•加拿大哈得孙湾
Accelerated Ice Sheet Breakup and Melting • Hudson Bay in Canada

----

图为哈得孙湾的冰盖碎裂与融化的状况。哈得孙湾位于加拿大北面的中部，是多雾、多冰、近封闭的亚北极内陆浅海，大部分海域处于北纬60°南北。

这里是北极熊生活的重要区域。经研究发现，这里的北极熊因为觅食困难，在过去20年里，数量下降了大约1/4。可见，气候变化对于地球南北两极生物多样性的影响是巨大的，对于全球气候变化的影响也是巨大的。

The picture shows the ice sheet breaking and melting in Hudson Bay. Hudson Bay, located in the middle of northern Canada, is a foggy, icy, nearly enclosed, sub-Arctic inland shallow sea, with most of its waters located north and south of 60°N latitude.

It is an important niche in habited by polar bears. Studies have found that the population of polar bears living here has declined by about a quarter over the past 20 years because of their difficulties in foraging. It can be seen that climate change has a huge impact on the biodiversity of the North and South poles of the Earth, as well as on global climate change.

南极企鹅•南极半岛
Antarctic Penguins • Antarctic Peninsula

在冰天雪地南极生活的明星动物就是企鹅，大家比较熟悉的主要有王企鹅、阿德利企鹅、巴布亚企鹅等。图中是白眉企鹅。

南极冰盖是世界上最大的冰原，这里储存了地球70%左右的淡水。据科学家预测，一旦南极冰盖融化，海平面将会提升55米以上。从图中可以看出，南极大陆不同于北极，是冰雪覆盖的陆地，在很多地方冰川延伸进入了南大洋，从而形成漂浮在海上的冰架、冰川。因此，南极冰川的总面积要大于南极大陆的总面积。

The star animals inhabiting the icy and snowy Antarctic are penguins. We are more familiar with the king penguin (*Aptenodytes patagonicus*), Adelie penguin (*Pygoscelis adeliae*), gentoo penguin (*Pygoscelis papua*) and so on. Captured in the picture is a gentoo penguin.

The Antarctic Ice Sheet is the largest ice sheet in the world, storing about 70% of the Earth's fresh water. Scientists predict that once the Antarctic ice sheet melts, the sea level will rise by more than 55 meters. As can be seen from the picture, the Antarctic continent, unlike the Arctic, is an ice-and-snow-covered land, and in many places the glaciers extend into the Southern Ocean, thus forming ice shelves and glaciers floating on the sea. Therefore, the total area of Antarctic glaciers is larger than that of the Antarctic continent.

雪中天鹅港·山东荣成
Swan Port in Snow ·
Rongcheng City, Shandong

这是大雪纷飞中的渔港——山东荣成港，人们也愿意称它"天鹅港"，因为每年都有数千只天鹅到这里和人们一同度过海港漫长的冬天。

地球上的大部分地方，人们最喜爱的天空降水大概就是雪了，无论在北方还是南方，人们看见雪都欣喜若狂，尤其是初雪——"瑞雪兆丰年"，就是好兆头！降雪是地球大气水循环不可分割的重要环节。

This is the snowy fishing port of Rongcheng City in Shandong Province. It is also known as "Swan Port" because every year thousands of swans would come here to spend the long winter together with the local residents of the port.

In most parts of the Earth, snow would probably be people's favorite form of sky precipitation. Whether in northern or southern China, people are ecstatic when they see snow, especially when they see the first snow. There is a Chinese proverb "Ruixue Zhao Fengnian", which literally means that the timely winter snow is a good omen for a bumper harvest in the coming year. Snowfall is an integral part of the Earth's atmospheric water cycle.

东边日头西边雨•海南五指山
Sun in the East and Rain in the West • Wuzhi Mountains, Hainan

----

　　海南热带雨林国家公园为中国最大的热带雨林公园，五指山属热带季风气候，年降水量1800~2000毫米。民谣有"又出太阳又下雨，栽黄秧吃白米"，寓光雨充沛、物产丰富、生活无忧之意。

　　中国的热带雨林在海南岛面积最大、最完整，其他地区如云南南部、西藏东南部（墨脱）、台湾南部还有小部分热带雨林。热带雨林往往是地球上降雨量最大的地方。与降雪相比，降雨在大气水循环中是水量更大、更重要、更普遍发生的主要环节。

The National Park of Hainan Tropical Rainforest is the largest tropical rainforest park in China. Wuzhi Mountains have the tropical monsoon climate, with annual precipitation of 1,800–2,000mm. There is such a folk song: "It is raining while the sun is out, and late rice is planted for future rice harvest", which symbolizes that this place is full of abundant sunlight and rain, rich in resources and people here can live a prosperous and carefree life.

The tropical rainforest in Hainan Island is the largest and most complete tropical rainforest in China. There are also a small number of tropical rainforests in other regions such as southern Yunnan, southeastern Tibet (Motuo County), and southern Taiwan. Tropical rainforests tend to be the places with the most rainfall on the earth. Compared with snowfall, rainfall is the more frequently occurring phenomenonin the atmospheric water cycle that typically comes with higher volume of water and plays more important roles.

广纳百川·纳米比亚
Inclusive of All Rivers • Namibia

从这个画面中你可以看到，太阳里面有城镇码头的影子，远处、近处都有很多水鸟在飞翔，近处穿过太阳的是火烈鸟群，空气的透明度相当难得。

降雪、降雨使水分回归大地。千条江河归大海，大海接纳了来自大陆所有湿地的水，也接纳了来自南北极的融水。至此，地球大气水循环完成了一个圆满的周期，并继续周而复始地重复着。

In this picture, you can see a glimpse of the town pier in the sun and water birds flying in the distance and near. Nearby, the flamingos (Phoenicopteridae) are passing through the sun, and the transparency of the air is quite rare.

Snow and rain bring moisture back to the Earth. Thousands of rivers flow into the sea, and the sea receives water from all wetlands on the mainland, as well as melted-water from the North and South Poles. So far, the Earth's atmospheric water cycle has finished a complete cycle, and it continues to repeat over and over again.

2 热力效应 泛三极
The Thermal Effects The Triple Poles

讲大气水循环,离不开讲地球"三极"。南北极对于全球气候的影响已经成为世界的共识,而青藏高原对全球气候的重要影响正在被人们所认识。青藏高原的隆起,形成了平均海拔4000米以上的"超高原",西风环流被青藏高原劈开后形成南北两支气流,太阳辐射产生的"热效应",使流经它的环流和季风改变了流向和力度,产生了绕行的"绕流"和爬升的"爬流"等气流分支,南北两支气流再在中国华南地区上空汇集,合并后继续影响着东亚的大气环流。

"泛三极"的概念是当前正在进行的"中国第二次青藏高原综合科学考察"提出来的,尽管这次重大考察工作目前还没有完全结束,但是中国科学家关于青藏高原对全球气候变化有重大影响的研究已经越来越被国际社会所接受。中国科学家将南极、北极和青藏高原这个地球第三极对于全球气候的重大影响称为"泛三极"效应。

隆起的青藏高原不仅大大影响了自身的环境,影响了中国西北的气候,也影响了青藏高原东南部、云贵高原、长江中下游的大气环流及降水,影响了南亚、东亚、东南亚的气候,并且跨洋过海,"遥相关"南极、北极,这就是"泛三极"效应。

在《多样性的中国湿地》一书中,我提出了新的中国湿地区划,将青藏高原湿地作为第一块最特殊的地域单独划分出来,称之为"青藏高原湿地区",并且提出"中国湿地区划,紧紧抓住的是两个要素:一是气候带,二是海拔差",并突出强调青藏高原湿地情况"这里的湿地水文、气候的微小变动都会波及全国乃至亚洲大地,是我们要竭尽全力来保护的'中华水塔''亚洲水塔'"。现在看来,当时提出的思路无疑是正确的。

青藏高原湿地不仅对中国,而且对亚洲气候都有重大的影响,它的每一点变化都会对全球气候变化产生微妙的后果。

青藏高原像一堵高墙,把印度洋产生的暖湿气流很大一部分挡在了南亚次大陆,使南亚成为世界降水最多的地区之一,使它南亚的大部分地区都具热带季风气候。暖湿气流也在雅鲁藏布江水汽通道下部和横断山余脉的

南坡产生了高于北回归线不少纬度的热带雨林。

印度洋暖湿气流无力向北越过相当厚实的、平均海拔4000米以上的青藏高原,因此其北面的中国西北地区降水量极少,成为世界上干旱和极干旱的主要地区之一。

面对喜马拉雅山脉,西南季风携带暖湿气流部分爬升,少部分艰难越过,更多的气流只能绕道往东,在中国的江淮地区及朝鲜半岛、日本岛,与绕过青藏高原北部的季风汇合,冷暖空气交锋在夏季形成了"梅雨",使原来地球同纬度大都干旱的区域变成了雨量充沛之地和"鱼米之乡",同时影响了华北和华南的降雨。

青藏高原所产生的"热力作用"和"动力作用",还使其东面的横断山脉若干峡谷变成了干热河谷,形成了"倒置的垂直地带性"植被现象。

这些都是青藏高原影响南亚、东亚、东南亚气候的特殊现象与异常表现。这种影响甚至跨越大洋到美洲。这就是为什么科学家把青藏高原与南极、北极一起影响全球气候的现象称为"泛三极"现象的原因。

A discussion about the atmospheric water cycle would be incomplete without a detailed look at the "three poles" of the Earth. The influence of the North and the South Poles on the global climate has become a global consensus, whereas the important influence of the Qinghai-Tibet Plateau on the global climate is only beginning to be appreciated. The uplift of the Qinghai-Tibet Plateau has formed a "super plateau" with an average altitude of more than 4,000 meters. The westerly circulation is split by the Qinghai-Tibet Plateau into two air currents—the north and south currents. The "thermal effects" generated by solar radiation changes the flow direction and strength of the circulation and monsoon that flow across the Plateau, resulting in the "winding current" and "climbing current", and other branches of airflow. The two air currents converge over South China and continue to influence the atmospheric circulation over East Asia.They meet their partner now and at some point in their journey of life. Azollaceae, also known as "red duckweed", is a fern that can carry out photosynthesis. I saw two different colors of Azollaceae floating on the water surface of the Yangtze River. They are like two lovers who meet by chance. Azollaceae grows in water and grows up in the moisture of sunshine, air and water. They meet their partner now and at some point in their journey of life. Azollaceae, also known as "red duckweed", is a fern that can carry out photosynthesis.

The concept of "the Triple Poles" was put forward by the ongoing "China's Second Tibetan Plateau Expedition and Research" project. Although this major expedition has not yet been fully completed, the research of Chinese scientists on the significant impact of the Tibetan Plateau on global climate change has been increasingly accepted by the international community. Chinese scientists refer to the significant impact of the South Pole, the North Pole and the Tibetan Plateau—the third pole on the Earth, on global climate as "the Triple-Poles Effect".

The uplifted Qinghai-Tibet Plateau not only greatly affects its own environment and the climate of northwest China, but also affects the atmospheric circulation and precipitation in the southeast of the Qinghai-Tibet Plateau, the Yunnan-Guizhou Plateau, the middle and lower reaches of the Yangtze River, and the climate of South Asia, East Asia, and Southeast Asia. It can even move cross the oceans to affect the South Pole and the North Pole, known popularly as "the Triple-Poles Effect".

In my previous book, *Diverse Wetlands of China*, I proposed a new zoning scheme for China's wetlands, singling out the wetlands of the Tibetan Plateau as the first and most special area, calling it the "Tibetan Plateau Wetland Area". I proposed that "the zoning of China's wetlands is firmly based on two elements: one is the climate zone and the other is the altitude difference" and emphasized the situation of the wetlands on the Qinghai-Tibet Plateau where "the small changes in the hydrology and climate of the wetlands here will affect the whole country and even the whole of Asia. It is the "Chinese Water Tower and 'Asian Water Tower' that we should do our best to protect". It now appears that the idea proposed at that time was undoubtedly correct.

The Qinghai-Tibet Plateau wetlands have a significant impact not only on China's climate, but also on Asia's climate, and every slight change of it would have subtle but non-negligible consequences on global climate change.

The Qinghai-Tibet Plateau acts like a high wall, blocking a large part of the warm and humid airflow generated by the Indian Ocean to the South Asian Subcontinent, making South Asia one of the regions with the most rainfall in the world, and making most parts of South Asia have a tropical monsoon climate. The warm and humid airflow also creates tropical rainforests at places whose latitudes is much higher than the Tropic of Cancer at the lower part of the water vapor channel of the Yarlung Zangbo River and the southern slope of the Hengduan Mountains.

The warm and humid airflow from the Indian Ocean is unable to cross northward over the rather thick Qinghai-Tibet Plateau with an average altitude of over 4,000 meters, so northwest China, which sits to its north, receives very little precipitation and is therefore turned into one of the major arid and extremely arid regions in the world.

Facing the Himalayas, the southwest monsoon climbs with warm and humid airflow, but only a small part of it can manage to climb over the mountain, and more of the airflow can only detour eastward. These airflows converge with the monsoon that bypasses the northern part of the Qinghai Tibetan Plateau in the Yangtze-Huaihe Region of China, the Korean Peninsula, and the Japanese Islands. The collision of cold and warm air forms the "Plum Rain" in summer, turning the original arid areas at the same latitude on the Earth into the land of abundant rainfall and the "land of fish and rice", and affecting the rainfall in North and South China.

The thermal and dynamic effects of the Qinghai-Tibet Plateau have also transformed several canyons of the Hengduan Mountains to the east of the Qinghai-Tibet Plateau into dry-hot valleys, creating "inverted vertical zonation" vegetation.

These are the special phenomena and anomalies of the Qinghai-Tibet Plateau affecting the climate of South Asia, East Asia, and Southeast Asia, which can also cross the oceans to affect the Americas. This is why scientists call the phenomenon of the Qinghai-Tibet Plateau, together with the South Pole and the North Pole, influencing the global climate the "Triple Poles" phenomenon.

雪岭冰川•新疆喀喇昆仑山
An Awesome View of Glaciers •
Karakoram Mountains, Xinjiang

　　青藏高原最西部的喀喇昆仑山是世界上高海拔山峰和最长、最美冰川密集分布的地方。

　　科学研究证明，印巴地区以及喀喇昆仑山降雨所释放的热量通过西风往北绕过青藏高原后，被输送到北极，影响北极气候，与北极"遥相关"；又往南绕过青藏高原，形成了暖空气对于南亚、东南亚的输送，从而影响南太平洋诸岛，甚至南极，又与南极"遥相关"。地球三极相互"遥相关"，共同推动了全球的气候变化。

　　The Karakoram Mountains, the westernmost part of the Qinghai-Tibet Plateau, are the world's highest-altitude peaks where long glaciers are most beautiful and most densely distributed.

　　Scientific research has proved that the heat released by rainfalls in the Indo-Pakistani region and the Mountains is transported to the Arctic after bypassing the Qinghai-Tibet Plateau through the westerly wind to the north, affecting the Arctic climate and making the latter "remotely connected" with the Arctic; and it also bypasses the Qinghai-Tibet Plateau to the south, forming a warm air that is transported all the way to South Asia and Southeast Asia, thus affecting the South Pacific islands and even the Antarctica, and making the Plateau "remotely connected" with the Antarctica. The three "remotely connected" poles of the Earth jointly produce a major impact onthe changes in global climate.

立体气候景观 • 尼泊尔博卡拉
The Vertical Climate Landscape • Pokhara, Nepal

　　在近7000米海拔的鱼尾峰下，生长在南亚热带的植物——花形奇特、色彩艳丽的三角梅恣意开放，与喜马拉雅山脉的皑皑雪山形成强烈的对比。

　　这是青藏高原南缘的喜马拉雅山脉将印度洋暖湿气流大量阻挡在南亚次大陆后产生的典型立体气候景观。

　　At the foot of the 7,000-metre-tall Yuwei Peak, the bright-colored bougainvillea—an exotic planted that is normally found only in tropical areas in South Asia—is in full bloom, contrasting sharply with the snow-capped mountains of the Himalayas.

　　This is a typical vertical climatic landscape that comes into being when the warm and humid airflow originated from the Indian Ocean is blocked at the South Asian subcontinent by the Himalayas that stand loftily along the southern edge of the Qinghai-Tibet Plateau.

高纬度热带雨林•云南盈江铜壁关
High-latitude Tropical Rainforest • Tongbiguan Pass, Yingjiang, Yunnan

----

这是中国最大的一片龙脑香科植物集聚区域，在高于北回归线1个纬度的地方出现这片典型的热带雨林实为罕见。这片区域地处南亚热带向北热带过渡地区，北部有高原屏障，南部受印度洋季风和印度半岛气团交替影响。复杂的地形和气候造就了其独特的热带雨林景观。

在中国还有一片更高纬度的热带雨林在西藏墨脱，那里居然高于北回归线6个纬度。这些都是青藏高原带来的特殊生态现象。

This is China's largest colony of Dipterocarpaceae plants, a typical tropical rainforest that can rarely be found to exist at a latitude that is one degree north to the Tropic of Cancer. Located at the transitional zone between the southern tropical belt in and the subtropical belt of the north, this region is sheltered in the north by the high-rising plateau and subjected in the south to the alternate influences of the Indian Ocean monsoon and the air mass originated from the Indian Peninsula. The complex terrain and climate work together to bring into existence this unique rainforest landscape here.

The other extensive stretch of high-latitude tropical rainforests in China is found in Metuo, Tibet, which sits up to 6 degrees north to the Tropic of Cancer. All these are the special ecological phenomena brought about by the Qinghai-Tibet Plateau.

干热河谷•云南怒江河谷
Dry and Hot River Valley •
Nu River Valley, Yunnan

　　图中的木棉花又称攀枝花,是干热河谷中常见的树种。干热河谷植被呈"倒置的垂直地带性"。

　　在青藏高原东面的横断山区,受季风影响的金沙江、怒江、澜沧江、元江等峡谷下方,出现的是具备干、热两个基本属性的带状区域,被称为"干热河谷"。这种气候是特殊的地形地貌形成的奇特气候,最主要的影响因素被归结为"焚风效应"和"山谷风局地环流效应"。

Kapok (*Bombax malabaricum*), also known as Panzhihua, is a tree species that is found quite commonly in dry and hot river valleys, where the vegetation typically appear in "inverted vertical zonality".

In the lower valleys of the Jinsha River, Nu River, Lancang River and Yuanjiang River that zigzag among the Hengduan Mountains on the east side of the Qinghai-Tibet Plateau and are subjected to the influences of monsoon climate, strips of lands characterized by extreme drought and high temperature often appear, which are popularly known as the "dry and hot river valleys". This is a peculiar climate caused by the special topography and landforms in local areas, influenced primarily by what are referred to as the "incineration wind effects" and the "valley wind local circulation effects".

阴雨天的徽州乡村•江西婺源
Rural Huizhou on a Rainy Day • Wuyuan, Jiangxi

图为由"山水林田湖草"诸多元素组合的婺源农村景观，此时这里雾气弥漫，很快就要进入梅雨季节了。

在北纬30度左右的环地球带上，世界的大部分地方都是荒漠。受青藏高原隆起的影响，处于这个环带的我国江南地区却雨量充沛，成了名副其实的"鱼米之乡"。每年6~7月的夏季风带来的降雨，使这里出现持续的天阴雨频现象，被称为"梅雨"。梅雨主要发生在我国的江淮及以南地区，以及韩国南部、日本南部等地。

The picture shows the rural landscape of Wuyuan, where a wide array of landscape elements like "mountains, forests, crop field, lake and grasses" are brought into harmonious co-existence. As it is drawing close to the plum rain season, a thin layer of mist can be seen to permeate in the air.

Most parts of the world that are situated over the belt encircling the Earth at around 30 degrees are deserts. However, thanks to the influence of the elevated Qinghai-Tibet Plateau, the Jiangnan region of China that happens to fall within this belt also is surprisingly blessed with abundant rain falls that make this part of the country a "land of fish and rice". As a result of the frequent rainfalls during the June through July period each year, the weather in this area at this time is often gloomy and rainy, known among local folks as the "plum rain". The influences of plum rain can mainly be felt in regions south to the Yangtze River in China as well as in certain southern parts of South Korea and Japan.

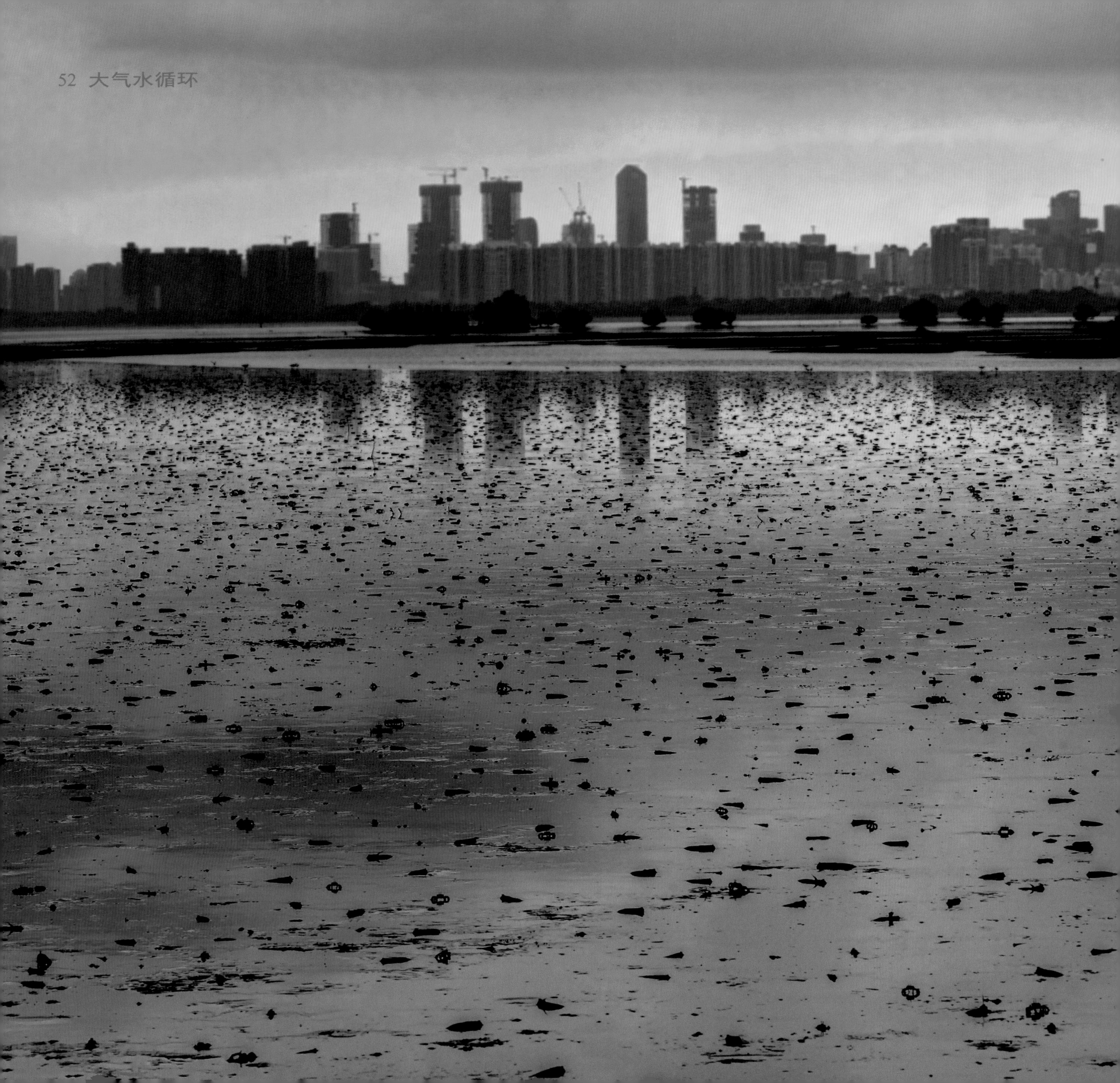

南亚热带海滨 • 香港后海湾
Coast in the Southern Subtropical Zone • Hau Hoi Wan (AKA the Deep Bay), Hong Kong

---

中国亚热带与热带分界线的大部分地区都是低于北回归线的，例如，广州、香港等，尽管都在北回归线以南，可冬季的最低温度常常低于2摄氏度，并不符合热带划分的条件。图为香港的南亚热带滨海滩涂湿地退潮时的景观，上面星星点点的是弹涂鱼和螃蟹。

青藏高原隆起及冬季西伯利亚寒流对中国大陆的作用，使中国东南部热带普遍南移，西南部北回归线以北部分区域出现热带雨林，这种奇怪的"满拧"现象主要是青藏高原带来的。

Most parts of the tropical and subtropical zones in China are located at places south to the Tropic of Cancer. Take for example, even though Guangzhou, Hong Kong and some other cities lie to the south of the Tropic of Cancer, the minimum temperature there often drops below 2 degrees Celsius in winter, making them not meeting the conditions for being categorized as tropical zones. Captured in the photo is the scene when the subtropical coastal wetlands in Hong Kong at the time of low tides, strewn on which are mudskippers and crabs that have been left behind the waves.

Due to the combined influences of the elevated Qinghai-Tibet Plateau and the Siberian cold current in winter, the tropical zones in Chinese mainland are on the whole slightly skewed southward. The unusual presence of tropical rainforests to the north of the Tropic of Cancer in certain parts of southeast China is caused by the presence of the Qinghai-Tibet Plateau.

荒漠之大成・甘肃阿克赛
A Masterpiece of the Desert • Aksai, Gansu

石山、岩漠、砾漠、沙漠、河流、乔木、灌木、沙地、草地及河流湿地，中国干旱地区荒漠的所有元素几乎都集中在这个视野里了。

这是我国西北干旱和极干旱地区荒漠之大成。

由于青藏高原对于海洋暖湿气流的阻挡，远离海洋的中国西北地区降水量极少且蒸发量极大，从而成为地球上最干旱的地区之一。

Almost all the composing elements of deserts in the arid regions of China, including rocky mountains, rocky deserts, gravel deserts, deserts, rivers, trees, shrubs, sandy lands, meadows and river wetlands, are captured in this photo. This explains why this place is often deemed to be the archetype and masterpiece of deserts in the arid and extremely arid areas of northwest China.

Blocked by the Qinghai-Tibet Plateau, the warm and humid airflows generated from the

oceans are unable to make their ways to faraway hinterland of northwest China where precipitation is extremely scarce and evaporation is high, hence reducing it into one of the driest regions on the planet Earth.

3 亚洲水塔 江河源
The Water Tower of Asia　The Source of Major Rivers

在青藏高原雪线以上，处处是雪山冰川，这里的冰川总储量占全中国冰川总储量的80%，冰川资源仅次于南极、北极，位于世界第三位，被称为"世界山地冰川王国"。

青藏高原湿地分布着数量最多的高原内陆湖群，湖泊星罗棋布，总面积达3万多平方千米，约占全国湖泊总面积的46%，是世界海拔最高、密度最大的湖泊群。

这里是亚洲多条大江大河的总发源地。青藏高原提供了黄河水量的1/2，长江水量的1/4，澜沧江水量的1/10。除了大家熟悉的世界第三长河——长江、世界第五长河——黄河两条大河全程都在中国境内外，亚洲大河湄公河上游的澜沧江，萨尔温江上游的怒江，伊洛瓦底江上游的独龙江，布拉马普特拉河上游的雅鲁藏布江，以及南亚大河的恒河、印度河的上游等，都是在青藏高原发源而成为国际河流的，这些大河滋润着南亚、东亚、东南亚的大片土地，涉及20亿人口的生活。因此，人们将青藏高原称为"亚洲水塔"。

这里产生的湿地是极为特殊的超高原（平均海拔4000米以上）高寒湿地，有着与世界上其他湿地都不一样的特殊性。据测算，青藏高原水资源量约为5.7万亿立方米，占全中国水资源总量的20%，是保障中国乃至南亚、东亚、东南亚国家水资源安全的重要生态高地。

Snow-capped mountains and glaciers abound on the Qinghai-Tibet Plateau in places whose altitude rises above the snow line. Accounting for over 80% of the country's total, the number of glaciers here ranks the third in the world, next only to that of the Antarctic and the Arctic, for which reason this place is hailed to be the "Kingdom of Mountain Glaciers in the World".

Wetlands over the Qinghai-Tibet Plateau are densely dotted with inland lakes and ponds whose total sizes add up to over 30,000 square kilometers, accounting for about 46% of the country's total lake areas. In other words, this is the highest part of the planet that comes with highest density lakes.

It is the place from which quite a few of the major rivers in Asia have their sources. The Qinghai-Tibet Plateau supplies 1/2 of the water volume of the Yellow River, 1/4 of that of the Yangtze River, and 1/10 of that of the Lancang River. Besides the Yangtze River (the world's third longest one) and the Yellow River (the world's fifth longest one) whose entire courses are within the territory of China, the Lancang River in the upper reaches of the Mekong River in Asia, the Nu River in the upper reaches of the Salween River, the Dulong River in the upper reaches of the Irrawaddy River, the Yarlung Zangbo River in the upper reaches of the Brahmaputra River, together with the Ganges and Indus Rivers in South Asia, all have their respective origins on the Qinghai-Tibet Plateau. These transnational rivers jointly nourish a vast area of lands in South Asia, East Asia, and Southeast Asia on which over 2 billion people live. For this reason, the Qinghai-Tibet Plateau is often honored as the "Water Tower of Asia".

Wetlands here belong to a very special type of alpine wetlands that are located on ultra-high places (averaging over 4,000 meters in altitude) and therefore bear some characteristics that no other wetlands in the world can compare. It is estimated the water resources storage on the Qinghai-Tibet Plateau is about 5.7 trillion cubic meters, accounting for 20% of China's total. In summary, it is an important ecological highland on which the water resources security of China and even South Asian, East Asian and Southeast Asian countries depend heavily.

白色游龙·青海昆仑山
A White Flying Dragon · The Kunlun Mountain, Qinghai

　　冰川如白色的游龙一般在皑皑众雪山中出没，或长或短，或粗或细，往往神龙见首不见尾。较高海拔处的冰川比较厚实饱满，较低海拔处的冰川细薄瘦小，在不同的海拔高度，冰川融化的程度不一样。从较低处的冰川我们可以清楚地看到，由于冰川的刨蚀作用而产生的长长的"U"形峡谷和冰川舌——这就是江河的源头。

　　中国是一个冰川大国，冰川总储量约为5590亿立方米，年平均冰川融水量为563亿立方米。

Countless glaciers, some long and some short, some thick and some thin, meander among the snow-capped mountains like white dragons that reveal themselves only occasionally to the earthly creatures like us. Glaciers at higher altitudes tend to be thicker and more robust, whereas those at lower elevations are comparatively thinner and smaller. Depending on the specific altitude at which they are located, the degree to which the glaciers melt varies notably. "U"-shaped canyons and glacial tongues created by glacier erosion—the source of rivers—are clearly visible at the low-lying glaciers.

China is a glacier-rich country, with a total reserve of approximately 559 billion cubic meters and an average annual glacier meltwater of 56.3 billion cubic meters.

南迦巴瓦峰云雾·西藏林芝
The Misty Nanga Bawa Peak • Nyingchi, Tibet

- -

　　在青藏高原面上，耸立着很多著名的雪山。图中是位于喜马拉雅山脉东端雅鲁藏布江大拐弯的南迦巴瓦峰，海拔7782米。

　　世界上最大、最深的峡谷——雅鲁藏布江大峡谷，成为印度洋暖湿气流北上最著名的"水汽通道"，充沛的热量和降水，使峡谷的下部产生了北半球最高纬度的热带雨林，还使峡谷的上部产生了世界上最高的树线（海拔4900多米），造就了此处特殊生态现象及独特景观。

Many famous snow-capped mountains stand toweringly over the Qinghai-Tibet Plateau. Shown in the photo is the 7,782-meter-tall Nanga Bawa Peak that stands at the great bend of the Brahmaputra River at the east tip of the Himalayas.

The Grand Canyon of the Yarlung Zangbo River, the largest and deepest among its

counterparts in the world, is home to the famous "water vapor channel" for the north-bound warm and humid airflow of the Indian Ocean. The abundant heat and precipitation that come with the airflow not only makes it possible for tropical rainforests to appear at the highest latitude in the northern hemisphere, but also brings into existence the highest tree-line (over 4,900 meters in altitude) at the upper part of the canyon, leading in result to the formation of a peculiar landscape and an ecological phenomenon here.

神山晨曦·云南香格里拉
The Sacred Mountain at Dawn • Shangri-La, Yunnan

- -

位于横断山脉中部怒江与澜沧江之间的梅里雪山，主峰卡瓦格博是云南省的最高峰，海拔有6740米，是藏族人民心目中神圣的神山之一，也是国家因文化保护而明令禁止攀登的唯一雪山。

南北走向的横断山脉和东西走向的喜马拉雅山脉、念青唐古拉山脉，合围形成的雅鲁藏布大峡谷"水汽通道"，是世界上水能资源最为富集的地区。这条通道缩小了南北自然带之间的明显差异，为许多生物物种提供了安全庇护，使这里成为全球生物多样性最丰富的区域之一。

Located between the Nu River and the Lancang River at the middle section of the Hengduan Mountains, the 6,740-meter Kawagebo Peak of the Meili snow-capped mountain is the highest peak in Yunnan Province. It is one of the most sacred mountains worshipped by the Tibetan people, as well as the only snow-capped mountain that is legally off-limits to mountaineers for the sake of cultural protection.

The North-South Hengduan Mountain Range, together with the East-West Himalayas and the Nianqing Tanggula Mountains Range, gives birth

to the world-famous "Water Vapor Channel" in the Grand Canyon of the Yarlung Zangbo River that has the richest hydropower resources in the world. This channel narrows down the gap between the natural belts in the north and the south and provides a safe haven for many biological species, making it rank among places in the world that have the richest biodiversity.

海拔最高的湿地•西藏珠穆朗玛峰
The Highest Wetland • Mount Qomolangma, Tibet

珠穆朗玛峰（图中远处的山峰）冰川融化流下的雪水小河——杂嘎曲，是世界上海拔最高的河流。

青藏高原湿地是世界上独具特色的湿地。这里由于常年冷冻低温、强烈的太阳辐射、冰雪融水、冻土交融、氧气稀缺、生物生长期极短，及不易氧化降解等特殊的水土光热组合等，使在这里生长的植物、动物、微生物以及由以上诸因素组成的生物本底都与其他地区有很大差异，形成全球最为独特的生态系统。

The Zagaqu River, which trickles initially down as melted water from the glaciers in and around Mount Qomolangma (the peak looming in the backdrop of the photo), is the highest river in the world.

The Qinghai-Tibet Plateau Wetland stands uniquely apart from its counterparts in the rest of the world. As a result of the combined effects of long-standing low temperature, strong solar radiation, ice- and snow-derived water, frozen soil, scarce oxygen, short growth span and not easy to be oxidized and degraded, the plants, animals, microorganisms that live here as well as the overall bio-system composed of these elements all differ remarkably from that of other regions, thus making it a very peculiar ecosystem on the Earth.

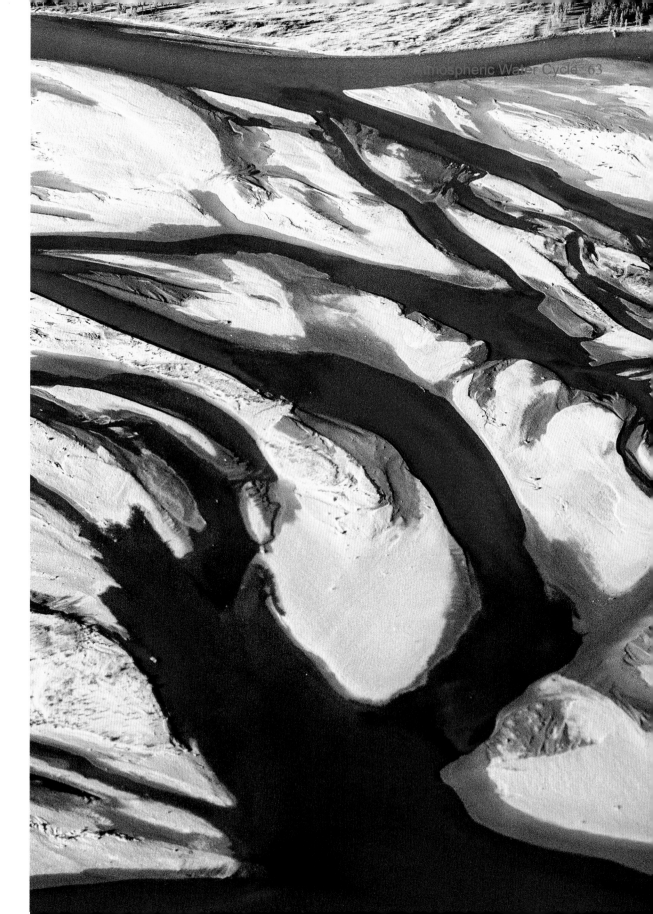

小溪汇成大河·西藏曲水
The Network of Streams that Feed into Mighty Rivers • The Zigzagging Streams, Tibet

----

　　无数冰川雪山融化形成的溪流最终汇成了大江大河，图为喜马拉雅山脉北部的大江——雅鲁藏布江的辫状河床。

　　受季节性冰川雪山融化的影响，雅鲁藏布江水荣枯变化比较明显，加上这是一条自由流淌的河流，因此，枯水时江滩裸露，河道纵横交错，颇具特色。

Countless tiny streams derived from melting glaciers and snowy mountains converge to give birth to mighty rivers. Shown in the photo is the braided-shaped riverbeds of the tributaries to the overwhelming Yarlung Zangbo River that runs across the northern part of the Himalayas.

The seasonal variation in the amount of water originated from glaciers and snowy mountains gives rise to significant changes in the volumes of the Yarlung Zangbo River during its water-rich and water-deficit phases. In dry seasons, the free-flowing tributary streams would often go dry, leaving on the plateau a complicated network of bare riverbeds for the viewers to marvel at the wonderful creation of Mother Earth.

千湖江河源·青海黄河源
The Lake-dotted River Sources • Source of the Yellow River, Qinghai

----

　　"地上的湖泊、天上的星"，是指青藏高原湖泊的数量犹如天上的星辰多得数也数不清。尤其是黄河源地区，远看星星点点、密密麻麻的都是湖泊、沼泽，这里存储了大量的水资源。青藏高原的湖泊数量占全中国湖泊总数的52%，是世界海拔最高的湖泊群。

　　青藏高原是长江、黄河、澜沧江、雅鲁藏布江等中国大江大河的发源地，也是亚洲许多大江大河的发源地。青藏高原水资源量约为5.7万亿立方米，是当之无愧的"中华水塔""亚洲水塔"。

　　"There are as many lakes on the Earth as there are stars in the sky" is a saying popularly used among the folks to describe the lakes that scatter densely across the Qinghai-Tibet Plateau. It is particularly the case at the place where the Yellow River has its source. Whichever way you cast your eyesight, you will see in the distance a rich cluster of lakes and marshes, in which immeasurable quantity of water resources are stored. Lakes on the Qinghai-Tibet Plateau add up to 52% of China's total, making it the highest lake group in the world.

　　The Qinghai-Tibet Plateau is the birthplace

of China's major rivers such as the Yangtze River, the Yellow River, the Lancang River and the Yarlung Zangbo River as well as the origin site of quite a few transnational rivers in Asia. Totaling approximately 5.7 trillion cubic meters in water storage, the Qinghai-Tibet Plateau is rightly credited as "the Water Tower of China" and "the Water Tower of Asia".

独特的高寒草甸沼泽·青海三江源
The Unique Alpine Swamp • Sanjiangyuan, Qinghai

在高寒高海拔气候条件下，生物种类组成较少，群落结构简单，氧化降解缓慢，并且只有一个短暂的生长期，这里的湿地是一种特殊的湿地，其他地方不可能产生如此大面积的此类湿地，这类生态系统被称为高寒湿地生态系统。

高寒湿地生态系统是指山地森林线以上到常年积雪带下限之间的，由适冰雪与耐寒旱的植物所组成的生态系统，包括高寒灌丛沼泽、高寒草甸沼泽、高寒垫状植被沼泽。

Under the cold high-altitude climate conditions where the bio-community typically consists of a mere few species, oxidizing and degradating slowly, that are of relatively short growth span, a very special type of wetland that no other places can give rise to would emerge. This type of wetlands is known as the alpine wetland ecosystem.

Situated normally between the tree-lines and lower-boundary of permanent snow belts in mountainous areas, alpine wetland ecosystem refers to the ecosystem composed of cold- and drought-tolerant plants that adapt well to ice and snow. This type of wetland ecosystem can be further divided into such subcategories as alpine shrub swamps, alpine meadow swamps, alpine cushion vegetation swamps.

山众曲多水富·青海囊谦
A Water-rich Realm of Mountains and Rivers • Nangqian, Qinghai

澜沧江是亚洲最大的国际河流，下游为穿越东南亚五个国家的湄公河。图为澜沧江流出青海囊谦县觉隆嘎峡时的景观。

澜沧江上游水系的扎曲、孜曲、巴曲、热曲、吉曲五条河贯穿囊谦县全境，全县人均自产水量达6.55万立方米，为世界人均水量的7.4倍，为我国人均水量的24倍。这个数字，在青藏高原颇具代表性。

The largest transnational river in Asia, the Lancang River is known in its lower reaches as the Mekong River that runs across five countries in southeast Asia. Shown in the photo is the scene the Lancang River rushes thunderously out of the Juelongga Gorge in Nangqian County, Qinghai.

Five rivers that contribute to the upper reaches of the Lancang River—namely the Zaqu, Ziqu, Baqu, Requ and Jiqu—run through Nangqian County, endowing local communities with an average per capita water resources of 65,500 cubic meters, which translates to 7.4 times that of the world and 24 times that of China. This figure is quite representative of the counties on the Qinghai-Tibet Plateau.

雪山下的冬冰湖·青海祁连
Frozen Lake at the Foot of Snow-capped Mountain · Qilian, Qinghai

----

在延绵的祁连山水的滋润下，干旱的河西走廊历来有"金张掖银武威"之说。张掖靠的是黑河水，武威靠的是石羊河水，疏勒河滋养了玉门，党河滋养了敦煌等，因此又有"一河养一城"之说。祁连山还孕育了黄河的支流庄浪河与大通河。

祁连山共有大小冰川2859条，总面积达1972.5平方千米，储水量811.2亿立方米，多年平均冰川融水量高达10亿立方米，发源于祁连山地的大小河流共有58条，祁连山是我国西北主要内流河的水源地，是名副其实的"西北水塔"。

Thanks to the nourishment of water originated from the rolling Qilian Mountains, two prosperous communities spring up in the arid Hexi Corridor, as known among local folks: "the Golden Zhangye and the Silver Wuwei". Just like that Zhangye owes its boom to the Heihe River, and Wuwei to the Shiyang River, Yumen and Dunhuang owe their success respectively to the Shule River and the Dang River nourishes. Hence comes the saying that "each city in the Hexi Corridor has its mother river to count on". The Qilian Mountains is also the place from which two chief tributaries of the Yellow River—the Zhuanglang River and the Datong River—are originated.

There are all together 2,859 large and small glaciers in the Qilian Mountains, which covers 1,972.5 square kilometers in its total area, boasts 81.12 billion cubic meters in water storage and generates annually up to 1 billion cubic meters of water on average that feed into 58 large and small rivers that have their origins here. Being the primary source of the many inland rivers in northwest China, Qilian Mountains indeed lives up to its honorable reputation as "the water tower of northwest China".

母亲河源•青海三江源
Source of the Mother River •
Sanjiangyuan, Qinghai

沱沱河为万里长江正源，发源于唐古拉山主峰的格拉丹东雪山的冰川中。图中高原面上河水自由地流淌着。

长江全长6380千米，是中国第一、世界第三大河流，也是世界上最长的完全在一国境内的河流，流域面积180万平方千米，约占全国土地总面积的1/5。长江和黄河一起并称为中华民族的"母亲河"。

The Tuotuo River—the genuine source of the Yangtze River—has its origin at the glaciers of Mount Gradangdong, the main peak of the Tanggula Mountains. Shown in the photo is the scene where the river flows leisurely across the plateau.

The 6,380-km-long Yangtze River ranks the first in China and the third in the world in length. It is also the longest river in the world whose entire course falls in entirety within the territory of a single country, covering in total 1.8 million square kilometers in its catchment, or approximately 1/5 of the county's total land area. The Yangtze River and the Yellow River are both credited as the "Mother Rivers" of the entire Chinese community.

湿地类型

在天空降水、地面径流、光热土和地形地势等综合环境因素的共同作用下，水汇流顺地势而下，成为了溪、河、江、滩及流域。水在地势的低凹处或山口等集水区上形成了各种湖泊；在河湖海边泥沙的不断淤积下形成了滩涂；在平缓洼地上与植物共同作用，氧化降解以及泥炭逐渐积累形成了不一样的沼泽；在入海河流和海岸潮汐的共同作用下形成了滨海湿地。

这些各类不同的湿地形成于不同纬度、不同海拔、不同光热土组合的环境中，使得依赖其生存的各种动植物种在其适合生长的环境中繁荣，最终构成了全球多样性的湿地生态系统。中国有《湿地公约》范畴内所有类型的湿地，还有世界独一无二的青藏高原特殊湿地，所以中国湿地生态系统的多样性和完整性是世界其他任何国家都无法比拟的。

Under the joint influences of different environmental factors such as precipitation, surface runoffs, light, heat, earth and geographical terrains, the water flows together down the slope and thus forms streams, rivers, beaches and river basins. Water come into shape in low-lying territories and mountain passes, and the consistent deposits of silts on the beds of rivers, lakes and seas finally bring into existence tidal mudflats. Through the combined actions of vegetation, oxygenolysis and the accumulation of peats, marshes would come into being in low-lying flats. And the combined effects of coastal tides and sea-bound rivers give rise to coastal wetlands.

Given that all these highly variegated wetlands come into being in different latitudes, altitudes and environments where light, heat and soil conditions vary significantly, a huge variety of vegetation and animals that adapt well to each setting would thrive, thus giving birth to the diverse wetland ecosystems of the planet. China's wetlands contain all the types of wetlands listed in the *Convention on Wetlands* and what's more, due to the presence of the unique wetland, the Qinghai-Tibet Plateau, Chinese wetland ecosystems are unparalleled throughout the world so far as their diversity and completeness are concerned.

WETLAND TYPES

1 河流湿地
Riverine Wetlands

按照全国土地分类和《湿地公约》分类方法，此部分主要包括河流、沟渠、运河以及瀑布、溶洞等。

According to the categories listed in the *Convention on Wetlands* and national land classification system, riverine wetlands includes rivers, ditches, canals, waterfalls and karst caves.

水土和声•甘肃张掖
A Chord of Water and Earth • Zhangye, Gansu

在干旱、半干旱气候条件下的祁连山北部，彩色砾石、砂岩和泥岩组成的山川，经长期风化剥离和流水侵蚀，形成了黑河流域这样独特的五彩地貌。这是水与土亲密合作的结果。

干旱、半干旱地区尚且如此，那么，在降水量较大的半干旱、半湿润地区，在降水量更大的湿润地区，水蚀的力量将会更加强烈地显示出来。水和土的共同作用，就能够给我们的世界带来希望、带来生命、带来繁荣。

As a result of thousands of years of weathering, exfoliation corrosion and erosion by flowing water under the arid and semi-arid climate in the northern area of the Qilian Mountain, mountains composed of gravels of colors, malmstone and mud rock gave birth to a unique landscape—the multicolored Heihe River Basin. It is a typical showcase in which water and earth worked jointly to create such a miraculous view of nature.

The phenomenon of water erosion can make such a difference in arid and semi-arid areas. The power of water erosion will be much more conspicuous in semi-humid and humid areas where the precipitations are higher. The constant interaction between water and soil is what bring our world hopes, lives and prosperity.

大地太极图 • 黑龙江南瓮河
The Tai Chi-pattern Land • Nanwenghe River, Heilongjiang

　　在中国北方的大兴安岭地区，河流蜿蜒曲折流淌在嫩江流域平缓的大地上，像一幅天然的太极图跃然画面上，河流的扭曲、摆动，把森林切成一块一块，这类形状的森林被称为"岛状森林"。

　　河流的这种现象，多半是发生在高纬度或高海拔有冻土的地方，由于有长年不化的冻土，加上地势平缓，河流下切不易而形成。

In the Daxinganling region of northern China, a river meanders through the flats of the Nenjiang River basin, shaping like a natural Tai Chi diagram. The twists and turns of the river cuts the forest into different parts and this kind of forest is called "Island Forest".

This kind of phenomenon of river happens typically in high-latitude and high-altitude areas because the soil there is kept frozen all year round, which, plus the flat terrain, makes it less likely for rivers to incise deep underneath.

帕隆藏布江云雾·西藏林芝
The Mist of Purlung Tsangbo River · Nyingchi, Tibet

　　帕隆藏布江水流湍急，云雾浓密，是给雅鲁藏布江流量贡献最大的支流之一。

　　雅鲁藏布江大拐弯是世界著名的印度洋暖湿气流的"水汽大通道"，陡峭并呈喇叭口的地形造成了大量的降雨，帕隆藏布江河流汇集的水量巨大，尤其在雨季，山洪暴发、两岸塌方频频发生。

The turbulent river crashes its way and dense mist envelopes the PurlungTsanbo River, the biggest tributary of the Yarlung Zangbo River.

　　The great Bend of the Yarlung Zangbo River is a "great channel of water vapour" for the world-famous warm and humid air currents of the Indian Ocean. The steep and bell-mouthed geographical structure brings about a high precipitation in this area. The area where the Yarlung Zangbo River converges has huge amount of water, especially in the rainy season when flashing floods and landslides occur with high frequencies.

怒吼吧，黄河·陕西壶口
The Thunderous Hukou Waterfall • Yellow River, Shaanxi

万里黄河从青藏高原一路走来，开山劈路，它沿途创造了若干的奇迹，养育了数千年来两岸的人民，却也背负了人们太多的希望和包袱，途经黄土高原就已经累了。黄土高原使黄河携带着大量的泥沙去给平原造陆，去奔向大海。

据资料记载，黄河最高含沙量时曾达每立方米920千克，曾经是世界上含沙量最大的河流。随着这些年来大有成效的造林绿化工程的实施，黄土高原的森林覆盖率已经达到59%，水土流失已经大大减少。

Beginning from the Qinghai-Tibet Plateau and flowing eastwards over thousands of mountains and hills, the Yellow River has over thousands of years created countless miracles in its marching while giving nourishments to generation after generation of people. However, it is also burdened with too many hopes and expectations of human being. It seems a little exhausted when passing the Loess Plateau. The Plateau adds huge amount of silt to the Yellow River, with which the mighty river is entrusted for the creation of vast lands on the plains as it makes its way towards the East Ocean.

According to statistics, the highest sand content of the Yellow River used to reach 920 kilograms per cubic meter, making it the river that had the highest sand content in the whole world back at that time. With the effective implementation of the massive afforestation projects, the coverage rate of vegetation in the Loess Plateau reaches 59% and the water loss and soil erosion also decrease greatly.

万里长江第一湾·云南香格里拉
The First Bay of the Yangtze River · Shangri-La, Yunnan

- -

长江、澜沧江、怒江从青藏高原奔流而下，穿越在高黎贡山、怒山和云岭等崇山峻岭之间，"三江并流"170多千米，在成为世界奇观留给世人后，长江至此拐了一个180度大弯向东北方向流去，并一直向东从上海进入太平洋。

假如没有这个万里长江第一湾，也许富庶的长江中下游平原将不复存在，中华民族的历史也将改写。

After plunging down from the Qinghai-Tibet Plateau, the three rivers—the Yangtze River, the Lancang River and the Nujiang River—meander side by side among the Gaoligong Mountain, the Lushan and Yunling mountains for more than 170 kilometers, impressing the world with a spectacular view of nature. Then suddenly, the Yangtze River takes a 180 degree turn towards the northeast direction and begins to flow eastwards

until it reaches Shanghai before flowing into the Pacific Ocean.

But for the existence of this first bay, the prosperous middle and lower reaches of the Yangtze River Plain would not have come into being and the Chinese history would probably have been rewritten.

跨国瀑布·广西崇左
Transnational Waterfall · Chongzuo, Guangxi

- -

德天瀑布位于中国与越南交界处的归春河上，与紧邻的越南板约瀑布相连，是亚洲第一、世界第四大跨国瀑布，瀑布气势磅礴、跌宕起伏，蔚为壮观，年平均水流量约为贵州黄果树瀑布的三倍。

世界四大跨国瀑布中前三名分别是：巴西和阿根廷之间交界处的伊瓜苏大瀑布、赞比亚和津巴布韦交界处的维多利亚瀑布以及美国和加拿大交界处的尼亚加拉瀑布。

Detian Waterfall is located at the Guichun River, the borderline between China and Vietnam, forming an integrated with the Ban Gioc-Detian waterfall situated in Vietnam. It is the greatest transnational waterfall in Asia and the fourth largest one in the whole world. The waterfall is of great momentum and everyone who has seen it would marvel at its mightiness and power. The annual water flow is about three times that of the Huangguoshu Waterfall in Guizhou Province.

The top three largest transnational waterfalls of the world are respectively the Iguazu Falls located at the border between Brazil and Argentina, the Victoria Falls bordering Zambia and Zimbabwe, and the Nigayara Falls bordering the USA and Canada.

岩溶地下河•广西百色
Karst Subterranean River • Baise,
Guangxi

- -

在喀斯特地区，水流或在地下，或溢出地表，形成了复杂的洞穴，产生了岩溶泉、断头河、地下河等特殊的湿地形式。

按照《第三次全国湿地资源调查技术规程》，"喀斯特溶洞湿地"归入"河流湿地"类。

In karst region, water is either under the ground or overflows the surface and thus forms several special wetlands such as complex caves, karst springs, beheaded rivers and subterranean rivers.

According to the *Technical Codes for the Third National Survey of Wetland Resources,* karst caves wetlands belong to river wetlands.

2 湖泊湿地
Lake Wetlands

按照全国土地分类和《湿地公约》分类方法，此部分主要包括湖泊、内陆滩涂、坑塘和盐沼、水库等。

According to the *Convention on Wetlands* and China's national land classification system, lake wetlands mainly include lakes, inland mud flats, pit-ponds, salt-marshes and water reservoirs.

青藏高原的湖泊·青海三江源
Lakes of the Qinghai-Tibet Plateau • The Source of the Three Rivers, Qinghai

----

　　高寒且平缓的青藏高原面上湖泊星罗棋布。这里是中国湖泊密度最大的地区，是青藏高原特殊湿地的代表区域，是在平均海拔4000米以上高原面上产生的。

　　处于世界屋脊上的湿地，是在低纬度高海拔高寒的环境下产生的，多为冰雪融水补给方式形成，较少受到人类活动的干扰。它的自然环境条件以及生成演化进程和其他湿地不一样，需要我们花大力气去专门研究和探索。

Large and small lakes are densely scattered in the high-altitude, cold and flat Qinghai-Tibet Plateau, which brings into existence a region with the greatest density of lakes. The region takes shape in the highland surface about 4,000 meters above the average sea level and is considered as the representative of special wetlands in the Qinghai-Tibet Plateau.

The wetlands located on the roof of the world are formed in the low-latitude, high sea-level and extreme cold conditions. Most of them come from melted snow water and have been subjected to relatively less interference by human activities. The unique processes through which they are created and evolve and the unique natural environment in which they sit render them significantly different from other types of wetlands. Therefore, we need to pay more efforts to study these wetlands.

雪山湖泊·西藏嘎隆拉山
Snow Mountain Lakes • Galongla Mountain, Tibet

　　这是喜马拉雅山脉南坡的高山湖泊群，湖水由冰雪融水及大量降水组成，由于湖水深浅及矿物质含量的不同，三个湖泊呈现出了不同的颜色。

　　这是高海拔高山、高坡上产生的湿地，它与青藏高原面上产生的湖泊的最大不同之处在于：降雨、降雪是此类湖泊水的主要来源，冰川融水只是补充。

　　This is a string of alpine lakes situated at the southern slope of the Himalayas. The lake water is composed of melted ice and snow water and abundant precipitations. Due to the different depth and mineral content, the three lakes vary from each other notably in colors.

　　This wetland is produced on high-altitude mountains and high slopes. The biggest difference between these lakes and their counterparts situated on the surface of the Qinghai-Tibet Plateau is that rainfall and melted snow are the main source of water for such lakes, while melted water from the glacial is only a complementary source.

长白山天池 • 吉林长白山
Changbaishan Tianchi • Changbaishan Mountain, Jilin

　　长白山天池是中国最大的火山湖，也是最深的高山湖泊，降雨、降雪是长白山天池水的主要来源。

　　长白山天池为公元1702年火山喷发后的火山口积水而成，高踞于长白山主峰的白头山（海拔2691米）为东北地区最高峰。

　　Changbaishan Tianchi is China's biggest volcanic lake and also the deepest alpine lake. The water mainly comes from the rain and melted snow.

　　The Tianchi Lake of Changbaishan Mountain is formed from the crater water after the volcanic eruption in 1702 AD, setting itself above the main peak (2,691 meters above the sea level) of the Changbaishan Mountains, Baitou Mountain, which is the highest peak in northeastern China.

正向演替·黑龙江齐齐哈尔
Ecological Succession· Tsitsihar, Heilongjiang

河流在平原上蜿蜒流动并进入洼地形成浅水湖泊，而随着湖泊逐渐长出植物，植物死亡不断更替变为泥炭，湖泊就逐渐开始淤积并沼泽化了，淤积越来越严重，湖泊面积也就越来越小。这种趋势表明湖泊走向了晚年，走上了消亡之路。

湖泊到沼泽的演替是湖泊成长、消亡的正常过程，是自然界的普遍规律。

Rivers wind through the plains, flow into the low-lying lands and then become shallow lakes. As time goes by, plants grow up in the lakes. When plants die and turn successively into peat, the accumulation of peat will make the lake develop into the marsh. The more severe the sedimentation, the smaller the lake will be. This phenomenon signals that the lake is in its senile years and on its way towards death.

The succession from lake to swamp is a normal process for lake's development to disappearance and is the universal rule of the whole nature.

红色的湖•内蒙古锡林郭勒
Red Lake • Xilingol, Inner Mongolia

由于卤虫体内的虾青素在高盐环境下呈红色，杜氏盐藻和嗜盐菌等在体内积累了大量的类胡萝卜素，这些微生物大量繁殖就会将湖水染成红色，所以盐湖并非没有生物。

在我国的西北干旱和半干旱地区，存在许多盐湖湿地，湖水含盐量多少不等。盐湖是湖泊在一定环境条件下发展到老年期的产物。

Astaxanthin in halworms will become red in high salt environment and there is large amount of carotenoids in the body of *Dunaliella salina* and halophilic bacteria. Once these microorganisms colonizes in massive quantities, the lake will become red. This bring us to the conclusion that salt lakes are by no means totally barren and lifeless.

In the arid and semi-arid areas of northwestern China, there are many salt lake wetlands, and the salt content of the lakes varies. The salt lake is a product of lake development when lakes get senile under certain environmental conditions.

浅水湖泊·江西鄱阳湖
Shallow Lake • Poyang Lake, Jiangxi

　　由于地势平缓的原因，长江中下游布满了许多大大小小的浅水湖泊。中国最大的湖泊——鄱阳湖就是这样一个大型的浅水湖泊。这些湖泊对于长江蓄洪补枯起着不可替代的作用，对于长江生态系统的重要性是无可替代的。

　　与洞庭湖等长江通江湖泊类似，浅水湖泊往往是雨季成湖、旱季为河，消长变化极大，旱季湖水消退就会出现大片的滩涂和草地。

　　Due to the flat terrain, the middle and lower reaches of the Yangtze River are covered with many shallow lakes, large and small. The Poyang Lake, the largest one in China, is just a large shallow lake. These lakes play an irreplaceable role in flood storage and replenishment of the Yangtze River and play an important role in stabilizing the Yangtze River ecosystem.

Similar to the Dongting Lake and other lakes that are connected directly with the Yangtze River, shallow lakes change significantly as season evolves. They become lakes in rainy seasons, and rivers in dry seasons. When water recedes excessively in dry seasons, a mass of mud flats and grasslands will come into being.

3 沼泽湿地
Swamp Wetlands

按照全国土地分类和《湿地公约》分类方法，此部分主要包括沼泽草地、森林沼泽、灌丛沼泽、沼泽地和盐沼、地热等。

According to the *Convention on Wetlands* and China's national land classification system, swamp wetlands include wet meadows, mangrove swamps, shrub swamps, swamplands, salt marshes and geo-thermal.

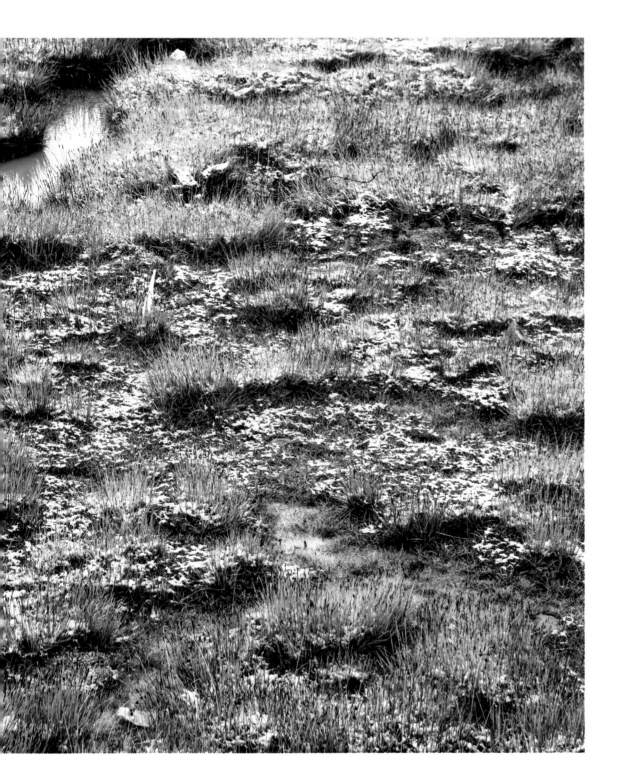

嵩草沼泽雪•青海三江源
The *Kobresia myosuroides* Swamp • The Source of Three Rivers, Qinghai

----

　　青藏高原的沼泽主要位于江河源区或近河源区，是在高寒气候、雪水冰水浸泡的特殊条件下形成的。

　　泥炭沼泽土、草甸沼泽土为这里的主要土壤类型，由于常年低温冷冻，这里的植物生长量非常少，植物残体分解特别慢，微生物也不活跃，与其他地区的沼泽有很大的不同。

　　The marshes of the Qinghai-Tibet Plateau mainly lie within or close to places where river shave their sources. They are typically formed in extreme cold weather and in the immersion of icy water and snow.

　　The main agrotype here are peat bog soil and meadow boggy soil. Because of the extreme coldness all year round, the vegetation coverage in these places is usually very low. The decomposition of plant residues is rather slow and the microorganism activities are not quite active, which differs greatly from swamps in other regions.

灌丛沼泽 • 四川康定
Shrub Swamp • Kangding, Sichuan

　　披上雪霜的灌丛，在深秋的沼泽草地上特别显眼，像一朵朵白花儿镶嵌在毛茸茸的红黄地毯之上。

　　以灌丛为主的淡水沼泽，其生存环境比森林沼泽要更加干旱残酷。

　　Shrubs covered with snowy frost are very conspicuous and eye-catching against the backdrop of the wet meadows in late autumn. Looked from afar, they are just like strain and strain of white flowers besetting in furry red-and -yellow blanket.

　　Freshwater swamps are mainly composed of shrubs and their living environment is usually much drier and severer than that of forest swamps.

大地锦绣·云南香格里拉
The Beauty of the Land • Shangri-La, Yunnan

- -

这个犹如地毯一般美丽的沼泽是云贵高原的沼泽，由于海拔的降低、光热条件的增加和降雨的增多，沼泽植物的生长和残物的分解都比青藏高原的沼泽要快得多，因此，沼泽草地显得更加厚实，色彩更加美丽。

香格里拉的海拔为3200米左右，相比青藏高原的沼泽，这里沼泽的生命力要强很多。

This carpet-like beautiful swamp is located in the Yunnan-Guizhou Plateau. As a result of the comparatively lower altitude, more favorable combination of light and heat, and more abundant precipitation, plants in these swamps are more flourishing than that of the Qinghai-Tibet Plateau and the decomposition activities are more active. Thus, wet meadows look more verdant and more beautiful in color.

The Shangri-La is about 3,200 meters above the sea level. Compared with swamps on the Qinghai-Tibet Plateau, the swamps here are more resilient and livelier.

森林沼泽•内蒙古根河
Forest Swamp • Gen River, Inner
Mongolia

　　兴安落叶松是东西伯利亚泰加林南缘的主要
优势树种，是寒温带针叶林的关键种。这种森林
同时具备山地植被和高山苔原带的特征，而冰沼
土和高位沼泽的出现就是其明显特点。图为兴安
落叶松林中的沼泽，这是冬季寒冷漫长，夏季温凉
多雨，常年季节性积水条件形成的。

　　森林沼泽主要分布在温带地区，在亚热带山
地和沿海地区也有零星分布。

Xing'an larch (*Larix gmelinii*) is the dominant
tree species on the southern edge of the Taiga
Forest in eastern Siberia, and it is a key species
in the cold temperate coniferous forest.This kind
of forest has the characteristics of mountain
vegetation and alpine tundra, and the presence
of ice marshes soil and highland marshes is its
distinctive feature. The picture depicts the marsh
dotted among the *Larix gmelini* forest in Xing'an,
which is formed under the conditions of cold and
long winter, cool and rainy summer, and perennial
seasonal ponding.

Forest marshes are mainly distributed in
temperate areas and also sporadically seen in
subtropical mountainous and coastal areas.

塔头沼泽·吉林延吉
Tatou Swamp • Yanji, Jilin

　　降雪给绿色、褐色的塔头地穿上了一件冬装，让塔头地更加紧密有致且突显身材。

　　塔头是一种特殊的沼泽湿地类型，特征是无数个高出水面几十厘米甚至一米的草墩连续不断, 这些草墩是由沼泽地里各种薹草的根系死亡后再生长、再腐烂，又再生长、再腐烂，周而复始，并和泥炭的长年累月紧密结合而形成的地貌。

　　Snowfall purifies the green and brown Tatou swamp and makes it tighter and more elegant.

　　Tatou is a special type of marsh wetland, which is characterized by the continual presence of grass mounds that extend tens of centimeters or even one meter above the water surface. These grass mounds are formed by endless cycle of the growth, decay, growth and decay of the roots of various bolting grasses in the marsh, which in turn cling tightly to peat through years of weathering and erosion to form this landform.

美妙的图形•甘肃阿克赛
Beautiful Pattern • Aksai, Gansu

红色的碱蓬呈六角形图案分布在白色的盐沼面上，绿色的盐生植物密密麻麻如满天星星，这个大苏干湖畔盐沼的奇异美丽景观真的耐人寻味。

地表水呈碱性，土壤中盐分含量较高，表层积累有可溶性盐，其上生长着盐生植物，是盐沼的基本特性。这片盐沼人若踩上去，表面就会伴随轻微破裂声下陷。在极干旱地区的苏干湖为山间断陷盆地集水而成，水源来自盆地东部大、小哈尔腾河，地表水通过小苏干湖（淡水湖）再注入大苏干湖（咸水湖）。

Hexagon-typed red seepweed (*Suaeda salsa*) scattered in the white surface of salt marsh, in which thousands of green halophyte grow, like starry night. The beautiful scenery on this lakeside of Dasugan Lake is really wonderful and inspiring.

The surface water is alkaline and the salt content in the soil is high. The soluble salt accumulates on the surface layer, over which halophytes plants grow. All these are the essential characteristics of salt marshes. If you step onto this marsh, the surface will sink down with a slightly cracking sound. The Sugan Lake, sitting in this extremely arid region, is formed from the standing water in faulted basin between the mountains that has its origin in the large and small Haiten Rivers situated on the eastern basin. The surface water flow through the small Sugan Lake (freshwater lake) and then injects into the large Sugan Lake (salt water lake).

热泉•云南腾冲
The Hot Spring • Tengchong, Yunnan

图为远观的腾冲温泉河。温泉河右边的蛤蟆泉是一个间歇热泉，由于温度很高，喷发出来即刻就变成了气雾状，并发出吼声。从图中可以看到，在如此高温环境下，周边植物也能够适应生长。

The picture shows the Tengchong Hot Spring River seen from afar. The Hama Spring located to the right of the Hot Spring River is a geyser. Due to the high temperature, the water spouting out will immediately be vaporized into vapors, accompanied with thunderous roars. We can see from the picture that the vegetation grows well even in such a high temperature environment.

湿地之大成·黑龙江虎林
A Masterpiece Wetland· Hulin, Heilongjiang

　　这是一片中国北方典型的大面积湿地，乌苏里江及其支流的经常性摆动，给这片土地留下了若干河道、牛轭湖、沼泽地以及沼泽草甸、灌丛沼泽、森林沼泽等，描绘出一幅中国东北湿地之大成图。

　　珍宝岛湿地国家级自然保护区是三江平原沼泽湿地集中分布区，是同纬度原始状态最具有代表性和类型最为典型的沼泽生态系统，一直按照其自身规律演替和发展，为亚洲东北部水禽迁徙的重要停歇地。

This is a typical large area of wetland in northern China. The frequent swing of the Wusuli River and its tributaries has left a number of riverways, oxbow lakes, marshes, marsh meadows, shrub marshes, and forest marshes to this land, presenting people with a great masterpiece of wetlands in northeastern China.

Zhenbao Island Wetland National Nature Reserve is an area in the Sanjiangyuan Plain where marsh wetlands are densely distributed. It is the most representative and typical marsh ecosystem in the primitive marshes of the same latitude. It has been evolving and developing according to its own laws, and is an important stopover site for migratory waterfowls in northeastern Asia.

4 滨海湿地
Coastal Wetlands

按照全国土地分类和《湿地公约》分类方法，此部分主要包括沿海滩涂、红树林和浅海水域、珊瑚礁等。

According to the land classification of China and the *Convention on Wetlands*, coastal wetlands mainly include coastal mudflats, mangroves (Rhizophoraceae) and shallow marine waters, coral reefs, *etc*.

退潮后的红树林·海南文昌
Mangroves in the Wake of Tide Ebb •
Wenchang, Hainan

　　这是具有热带、亚热带河口地区湿地生态系统典型特征的红树林带，红树林在海滩上形成了一道篱笆，发达的支柱根加速了淤泥的沉积作用，并促进陆地面积逐渐向外扩张。面对海潮风暴的袭击、海啸巨浪对海岸的破坏，有红树林保护的海岸就要安稳平和得多，因而红树林有"海岸卫士"之称。

　　红树林的咸淡水交叠的生态环境，为众多的鱼、虾、蟹、水禽和候鸟提供了栖息和觅食的场所，蕴藏着丰富的生物资源和物种多样性。

This is a mangrove (Rhizophoraceae) belt with typical characteristics of wetland ecosystem in tropical and subtropical estuaries, forming a fence on the beach. The highly developed prop roots of mangroves speed up the deposition of silts and gradually push the land area to expand outwards. Facing the attacks from tidal storms and the damages to the shore caused by tsunami waves, coasts protected by mangroves enjoy much more stable and peaceful time. Therefore, they are known as "the guardian of the coast".

Mangroves ecological environment where brackish and freshwater intermingles provides a favorite habitat of rest and foraging for numerous wildlife species, including fish, shrimp, crabs, waterfowl and migratory birds, forming a vital paradise with rich biological resources and diverse species.

红海滩·辽宁盘锦
The Red Beach • Panjin, Liaoning

在辽河口大面积的滩涂上，生长着翅碱蓬这种潮间带植物，这是大海与陆地的过渡，淡水和海水的交汇、沙土和盐碱的相互循环作用下形成的。这大片的碱蓬草，每年四五月开始由绿慢慢变红，到了九月便像落下的红霞在海陆间燃烧，显得格外奇特美丽。

在河流入海口，淡水携带大量的营养物质沉积在这里，并在潮水反复浸淹作用下，形成了适宜多种生物繁衍的河口湾湿地。这里是世界濒危鸟类——黑嘴鸥最大的繁殖地。

The large mudflat near the estuary of Liao River, which is formed by the intersection of fresh water and sea water, sand and alkali, is a transitional area between the sea and the land overgrown with *Suaeda salsa*, a kind of wing-shaped seepweed belonging to intertidal plants. This extensive stretch of seepweedon this beach would start to turn gradually from green to red over the period ranging between April and May each year, culminating in September into a splendid blanket of crimson haze that covers the entire zone between the land and the sea.

At the entrance of the sea, fresh water carries a large amount of nutrient to be deposited here and creates an estuarine wetland suitable for the reproduction of various creature safter repeated inundation by the tide. There is also the largest

breeding habitat for the black-billed gull (*Larus saundersi*), an endangered bird in theworld.

海蚀地貌·海南西沙东岛
Marine Abrasion Landform • The East Island of Xisha, Hainan

海蚀多发生在基岩海岸，图为西沙东岛老龙头珊瑚礁岩被严重海蚀的状况。

海蚀作用有三种：冲蚀、磨蚀与溶蚀。海蚀的程度与当地海浪的强度和海岸原始地形有关，也与组成海岸的岩性及地质构造特征有关。海蚀地貌有海蚀崖、海蚀台、海蚀穴、海蚀拱桥、海蚀柱等。

Marinea brasion occurs mostly on bedrock coasts. The picture shows the severe sea abrasion of Laolongtou coral reef rocks in the East Island of Xi sha.

There are three types of marine abrasion: erosion, abrasion and dissolution. The extent of marine abrasion is related not only to the strength of waves and the original topography, but also to the lithology and the geological structure of the coast. Marine abrasion can lead to the formation of cliffs, platforms, caves, arch bridges, columns and other kinds of landforms.

南海潟湖 • 海南西沙琛航岛
The Lagoon of the South China Sea • Chenhang Island in Xisha, Hainan

----

图为西沙群岛琛航岛的珊瑚潟湖。潟湖是被沙嘴、沙坝或珊瑚分割而与外海相分离的局部海水水域，海水较浅且相对平静，这里是海洋生物尤其是底栖动物的优良栖息场所。

南中国海上分布着无数的岛屿、沙洲。这些岛屿大部分是由珊瑚砂岩组成的，也伴随产生了很多潟湖。

The picture shows the coral lagoon of the Chenhang Island in Xisha. A lagoon is apartial body of relatively shallow and calm seawater that is separated from the deep sea by sand spits, sand dams or coral reefs. It makes up an ideal habitat for marine life, especially benthic animals to live in.

There are countless islands and sandbars in the South China Sea. Most of these islands are composed of coral sandstones, which also give rise to numerous lagoons.

湿地生物

依靠湿地生活的野生生物，我们可以举出大量的例子，例如，人们通常所说的水禽、水生动植物或者湿地物种等，而对于有些物种来说，我们却很难来确定其是否符合这个要求。如果仅仅采用水生野生生物或陆生野生生物来把它们区别开来，其实是一个不太科学的划分方法。

譬如，水禽包括游禽、涉禽等，我们可以把它们都简单地归为湿地动物，但是对于两栖动物和爬行动物有时就很难区分开了。是以其生活史中的大部分时间在水里还是在陆地上来衡量，还是以在水中生育或在陆地上生育来划定？科学家们对此也争论不休。即便是争议较少的兽类，有的就生活在水里，有的虽然生活在陆地但非常依赖湿地生活，也很难用简单的方法来把它们划分清楚。为了解决这个问题，有时候就只能用行政管理的手段来确定了。譬如，扬子鳄原是被划为陆生野生动物的，《国家重点保护野生动物名录》（2021版）将其划为了水生野生动物。

We can name a large number of examples of wild animals living in wetlands, such as waterfowl, aquatic animals and plants or wetland species that people are familiar with. However, for some species, it is difficult for us to determine whether they meet such standards. It actually is a less scientific categorization that distinguishes them merely on basis of the differences between aquatic wildlife and terrestrial wildlife.

For example, waterfowl, including natatorial bird (Natatores), wader, *etc*., can be simply categorized as wetland animals, but sometimes it is hard to distinguish amphibia from reptilia. There is still much disputes among scientists as to whether the dividing line between terrestrial and aquatic animals should be drawn on basis of the fact that the major part of their lifetime is spent in water or on lands, or on the fact that they are born in water or on lands. Even for those less controversial animals, it is not easy to draw a clear line, given that whereas some live in water purely, some others depend heavily on wetlandseven though they mainly live on lands. Sometimes, administrative measures will have to be resorted to for the solution of this problem. For example, Yangtze crocodile (*Alligator sinensis*) used to be categorized as a terrestrial wildlife until the *List of Key Protected Wild Animals in China 2021* changed it into aquatic wildlife.

1 湿地鸟类
Birds of Wetlands

　　湿地中物种最多、数量最大的野生动物种类就是鸟类了。《湿地公约》最初提出保护的，以及迄今一直强调要重点保护的都是"作为水禽栖息地的国际重要湿地"。水鸟（这里包括水禽）是湿地的精灵，是湿地生态系统生命和活力的要素体现。

Among all the wildlife inhabiting wetlands, it is birds that have the richest diversity in species and the largest populations. The *Convention on Wetlands* was initially put forward for the protection of the wetlands of international importance especially as waterfowl habitats, which is still regarded as the focus even up to today. Water birds (including waterfowl) are the elf of wetlands, showcasing the life and vitality of the ecosystem of wetlands.

万物生长靠太阳·青海青海湖
All Living Things Depend on the Sun for Survival · Qinghai Lake, Qinghai

红彤彤的太阳从青海湖东边升起来了，聚集在石头城堡顶上的鸬鹚苏醒了，伸伸脖子、理理翅膀，鸟群中出现了阵阵的躁动，它们正在准备迎接初升的太阳。新的一天开始了。

青海湖国家级自然保护区海西皮岛前有一巨石突兀嶙峋，矗立湖中。四周波光岚影，颇为壮观。巨石之上，鸬鹚窝一个连一个，像一座鸟儿的城堡。青海湖的普通鸬鹚数以万计，它们主要以青海湖里的鱼类为食。

The red sun is rising from the east of Qinghai Lake, awakening the cormorants (*Phalacrocorax carbo*) gathering around in the roof of stone castle. They stretch their necks and smooth their wings, making waves of stir among the flocks. They are waiting for the newborn sun, waiting for a new day to begin.

There is a singular huge rock standing firmly in the lake in front of Haixipi Island of Qinghai Lake National Nature Reserve, surrounded by the waves and shadows of the water, which is rather spectacular. The nests of cormorants are dotted everywhere like a bird castle on the rock. There are millions of cormorants in Qinghai Lake, feeding themselves primarily on the fishes in the lake.

日月同辉 • 河北衡水湖
The Sun and the Moon Shine Together • Hengshui Lake, Hebei

　　西边的太阳就要落山了，湛蓝的天空依然热闹非凡，鸟儿们飞来飞去。月亮已经悄悄升起，此时低角度的太阳光照亮了一对晚归大天鹅的腹部和翅膀下面，呈现出一幅自然界日月同辉的美丽画面。

　　衡水湖国家级自然保护区是华北平原面积较大，集水域、沼泽、滩涂、草甸和森林等生态系统为一体的湿地自然保护区之一，其物种资源有明显的古北界动物特点。

The sun is going down in the west, while the birds still fly happily, making the blue sky remain hustling and bustling. The moon has risen obliviously, when the sun at low angle shed light on the lower part of the bellies and wings of the two late-returning swans (*Cygnus Cygnus*), presenting a picturesque natural scene that the sun and the moon shine together.

Hengshui Lake National Nature Reserve is one example among the many large wetland natural reserves in North China Plain that combine ecosystems like water, swamps, tidal flats, meadows and forests into integrated wholes. The wildlife species resources here show obvious characteristic of the Palearctic animals.

向我飞来·河南黄河湿地
Flying up to Me • The Huanghe River Wetland, Henan

芦苇荡前，一群灰雁迎着阳光起飞而来。成群的大型鸟类面向镜头飞起的画面还是少见的，说明此时的它们丝毫没有感觉到摄影者的存在——我一点都没有干扰它们。

灰雁、鸿雁、豆雁、白额雁等雁属类都是大型迁徙鸟类，在中国的种群数量较大。飞行时成有序的队列，排成"一"字形或"人"字形等。中国文化中喜欢用它们来寄托季节变换和思乡的情结。

A flock of greylag gooses (*Anser anser*) are flying alongside the sunlight in front of reed marshes. It is a rare scene that flocks of large-size birds start to fly towards our cameras, which means that they completely do not feel the existence of photographers—we are not disturbing them.

The greylag goose, the swan goose (*Anser cygnoides*) the bean goose (*Anser fabalis*) the white-fronted goose (*Anser albifrons*) and other geese are large migratory birds with a large population in China. When flying, they form an orderly queue, and line up in a line, or in the form of Chinese character "人", *etc*. Their images are often used in Chinese culture as the symbols of seasonal changes and homesickness.

激情四射 • 吉林长白山
Energy and Passion • Changbai Mountains, Jilin

自然界里，雄性动物在追求爱情方面往往是激情四射、不遗余力的。中华秋沙鸭这种自然界里最古老的鸭子也毫不例外。

中华秋沙鸭主要分布在中国，栖息于阔叶林或针阔混交林里的溪流、河谷、草甸和水塘中。它们在溪流中交配，繁殖巢多筑在紧靠水边的大树洞中，洞口距地面往往10米左右，孵化出来的小鸭子必须自己从树洞中直接跳下，生存下来的小鸭子才能跟上妈妈一起去寻找食物。

In nature, male animals are always full of energy and passion in the pursuit of love without any regrets left. Chinese merganser (*Mergus squamatus*) is no exception.

Chinese mergansers are mainly distributed in China, inhabiting streams, valley rivers, meadows and ponds in broad-leave forests and coniferous and broad-leave mixed forests. They mate in streams and build breeding nests in tree holes nests next to the waters. The holes are usually 10 meters from the ground, from which the newly hatched ducklings must jump down by themselves. Only those little ducks which survive can follow their mother to look for food.

初沐阳光•青海青海湖
Enjoying the First Ray of Sunlight •
Qinghai Lake, Qinghai

- -

斑头雁的雏鸟从蛋壳里陆续出来了，趁妈妈暂时离开的短暂时间里，它们尽情地享受着人生的第一缕阳光，浑身暖意融融。新生命的世界该多么美好啊！

斑头雁属陆栖鸟，善行走，繁殖在高原湖泊周边，尤喜咸水湖。斑头雁尽管游泳很好，但多数时间都生活在陆地上。常见斑头雁与棕头鸥混群繁殖，亦见与黑颈鹤、赤麻鸭等鸟类混群，互不排斥。

Look! The chicks of bar-headed goose (*Anser indicus*) are coming out of eggshell one after another. Taking advantage of this short period time when their mother is away, they are enjoying the first ray of the sunlight in their lives as much as they want. What a nice world! They feel so warm.

Good at walking, the bar-headed goose is a terrestrial bird. They breed around the plateau lakes and saltwater lakes are their favorite. Though they can swim well, they spend most of their life time on the land.

It is common for bar-headed goose and brown-headed gull (*Larus brunnicephalus*) to breed in mixed groups as well as with black-necked crane (*Grus nigricollis*), ruddy shelduck (*Tadorna ferruginea*) and other birds.

红杏出墙·江西鄱阳湖
The Forlorn Lover • Poyang Lake, Jiangxi

　　鸟类的世界也有第三者，因而红杏出墙也就不奇怪了。这只雄性白鹭带着鱼回家来献殷勤，却发现已被第三者霸占了爱巢！这种事不常见。

　　白鹭是鹭类中体型较大的物种，鹭科的鸟是人类认识较早的鸟类之一，由于体态优美，成为古人诗歌中经常赞美的对象。白鹭繁殖期为每年4~7月，营巢于高大的树上或芦苇丛中，多集聚而营建群巢。

　　There are home wreckers in the world of birds, so one bird having affairs with another is not strange. This male egret come home with fish to please his wife, only to find that another male wrecker has forcibly occupied his home! This kind of events is not seen often.

　　Egret is a species of herons that typically comes with large size.Herons are one of those birds that were known by human earliest. The beautiful body shape makes them the objective of ancient people's praise in poems. The breeding period of egret is April to July each year. They construct nests in tall trees or reeds, and they always assemble and build nest where they live in groups.

晚渔 • 广东珠海
Fishing at Dusk • Zhuhai, Guangdong

- -

海浪在岩石中不停地涌动，忽退忽进，而岩鹭的眼睛紧紧盯着海浪中的鱼，它喜欢在海岸的岩石和海浪中跳来跳去，翅膀张开并不是要起飞，而是在无序的浪中寻找平衡。既抓住鱼又躲开浪，这才是真本事！

鹭科鸟类中岩鹭有点与众不同，一是颜色独特——全身呈炭灰色，这个颜色和海边岩石的颜色非常接近；二是不像其他鹭类在平静的水里捕食，而是在海浪中捕食，是典型的海岸鸟。这种鸟一度非常稀少，现在也不多。

The pacific reef heron (*Egretta sacra*) stares intently at the fish among the sea waves that break incessantly onto the rock. It enjoys jumping around seaside rocks and waves, opening wings not to fly but to keep balance in unpredictable waves. Being able to catch the fish without being caught in the wave is the trick that makes them impressive.

The pacific reef heron is a bit different from its other peers among the heron family. First, the color is unique — the whole body is charcoal gray, which is very close to the color of the seaside rocks. Second it is a typical coastal bird that hunts in the waves, unlike other herons that typically hunt in the sea. This type of birds used to be very rare, nor are they seen often today.

低调的飞行冠军•辽宁鸭绿江口
The Low- profile Flying Champion • the Outlet of Yalu River, Liaoning

----

　　谁也不可能想到，这个在海滩上独自安静觅食的小家伙，竟然是一位不间断飞行距离最长的世界冠军！

　　斑尾塍鹬可以从新西兰的北岛飞到中国的鸭绿江口，补充体力后再飞往美国的阿拉斯加，又从阿拉斯加直接飞回新西兰。它能够在中途不降落、不进食的情况下长时间飞行，行程最长可达1.1万多千米 。

Who could have imagined that this little bird that forages quietly alone on the beach is in fact the world's flying champion capable of completing the longest non-stop flight!

The bar-tailed godwit (*Limos alapponica*) could fly all the way from the North Island of New Zealand to the Yalu River of China without stop. After a brief stopover to top-up and replenish its energies, it will go on its flight to Alaska of the USA before heading back to New Zealand. It could fly for long time without stop and eating in the interval, covering a maximum distance of over 11 thousand kilometers.

南海之鹬•*南海永乐群岛*
Sandpiper (Scolopacidea) of the South
China Sea • Yongle Islands, the South
China Sea

- -

　　在远离大陆的西沙群岛羚羊礁上，一群翻
石鹬在上下翻飞。它们时而在沙滩上觅食，时
而在海面上飞翔。

　　翻石鹬喜欢在海岸海岛的潮间带、沼泽或
礁石等环境中寻找食物，主要以藏身石块或沙
砾下的沙蚕、螃蟹等小动物为食，故而得名。

　　鹬鸟是涉禽中种类最多、数量最大的一
类，飞行能力极强，是迁徙于世界各类湿地水
鸟的重要组成部分。

On the antelope reef of Xisha Islands far away
from the mainland, a flock of ruddy turnstones
(*Arenaria interpres*) flip up and down. Sometimes
they forage on the beach and sometimes fly above
the sea.

The ruddy turnstone likes to find food
in the intertidal zone, swaps, reefs and other
environments in coastal islands, feeding on clam

worms (*Nereis succinea*), crabs (*Brachyura*) and
other small animals hiding under the rocks or grits,
hence the name.

The sandpiper is a type of wading birds that
has the largest variety and the largest population.
It excels particularly in flying and is an important
component of migratory waterfowls in various
wetlands around the world.

水上凤凰•湖北武汉
Water-borne Phoenix • Wuhan, Hubei

- -

水雉步履轻盈，善于在莲、菱角等浮水植物上舞蹈，因体态优美，羽色艳丽，被称为"水上凤凰"。

水雉善于在浮水植物上行走是因为它长有一对又细又长的脚爪。不仅如此，它也善于游泳和潜水，在水中算是文武双全了，真不枉人们给它赋予了那么一个尊贵的名头。

With its light steps, the pheasant-tailed jacana (*Hydrophasianus chirurgus*) is good at dancing on lotuses, water chestnuts and other floating plants. It is called "water-borne phoenix" because of its beautiful color and elegance.

Its thin and long paws allow the pheasant-tailed jacana to walk on the floating plants. Besides, it is also good at swimming and diving, kind of versatile in the water and it totally deserves the honorable title that people give it.

守株待兔•青海青海湖
Waiting for the Windfalls • Qinghai Lake,
Qinghai

在青海湖附近生活的棕头鸥应该是世界上最幸福的棕头鸥了，这里食源充足，尤其是每年都要举行的一场饕餮盛宴，棕头鸥是从来都不会缺席的。

在裸鲤洄游的季节，湖内即将产卵的鱼相约成群溯流而上，直至找到合适的地方排卵受精。由于河道里鱼的密度很大，产生"半河清水半河鱼"的景象并不夸张。在此时段，棕头鸥大量集聚在河道附近"守株待兔"，抓捕河中的美食，会形成"鱼跳龙门""群鸟猎鱼"等现象，堪称世界奇观。

Brown-headed gulls (*Larus brunnicephalus*) living near Qinghai Lake could probably be regarded as the happiest brown-headed gulls in the world, because they have abundant food here so that they won't miss the grand banquet each year.

During the migratory season of naked carp, the fish that is to spawn would assemble to swim upstream until they manage to find a suitable place for ovulating and being fertilized. Due to the high density of fish in the river, it is not an exaggeration to say that "the river is half filled with clear water and half filled with fishes". During this period, a large number of brown-headed gulls would arrive in huge flock sat the river where they can easily prey on fishes teeming in the river. A miraculous show in which flocks of birds swoop down at the unsuspecting fishes for a heart feast would be put on stage.

晨出•黑龙江扎龙
The Sunrise • Zhalong, Heilongjiang

太阳初升，丹顶鹤起飞，逆光的丹顶鹤既不失剪影效果又有细节层次，姿态优美，物种辨识度很高。全世界的15种鹤中，中国分布有9种。

丹顶鹤在中国传统文化中具有崇高的地位，被称为"仙鹤"。明清时期一品文官服上才有资格绣上仙鹤图案。中华文化赋予丹顶鹤美丽、飘逸、长寿、吉祥和高贵的寓意，已经成为精神高洁的象征。

When the sun rises, the red-crowned crane (*Grus japonensis*) starts to fly. The red-crowned crane flying against the sun looks extremely elegant and beautiful, projecting into the camera lens a nice silhouette effect filled with details that make the bird easily recognizable. Among the 15 species of cranes in the world, 9 species are found in China.

The red-crowned crane is held in a very high position in traditional Chinese culture and is regarded as the "celestial crane". During the Ming and Qing Dynasties, only the first-grade officials were qualified to embroider patterns of the crane on their official robes. Chinese culture endows the red-crowned crane with the meanings of beauty, elegance, longevity, auspiciousness and nobility, and has become a symbol of spiritual purity.

冬奥会的吉祥鹤·北京野鸭湖
Lucky Crane of Olympic Winter Games · Yeya Lake, Beijing

2022北京冬奥会如期顺利举行，位于北京延庆的国家高山滑雪场迎来了冬奥会期间最大的一场雪，整个滑雪场和小海陀山都笼罩在一片皑皑白雪之中。这时，灰鹤列队展翅而来，穿过天空的瞬间与国家高山滑雪中心同框，形成一幅"鹤喜冬奥"的绝美画卷。

灰鹤的到来为冬日的延庆赛区增添一份灵动的同时，下雪不多的海陀山竟然也披上了厚厚的雪衣，它们的同时到来仿佛在预祝北京冬奥会一切顺利！

Olympic Winter Games Beijing 2022 unveiled as scheduled. The National Alpine Skiing Centre in Yanqing (Beijing)has ushered in the heaviest snow during the Olympics, with white snow covering the whole skiing center and Xiaohaituo Mountain. Grey cranes line up and spread their wings. At the moment their images are captured in my lens, they happen to be flying above the center, hence coming into being this fantastic view in which cranes joins human being in the happy celebration of the Olympics.

Apart from the vitality brought by the arrival of cranes, the Xiaohaituo Mountain which does not snow often is also wearing a thick snow coat. Their simultaneous arrivals look like an auspicious sign that foretells the smooth progress and success of the Beijing Winter Olympics.

白鹤·江西鄱阳湖
Snow crane • Poyang Lake, Jiangxi

清晨的湖面上浓雾刚刚散开，兴奋的白鹤就开始不断地鸣叫和舞蹈，浓雾将所有的背景都遮挡住了，仅白鹤显露其中，呈现给我们一幅十分雅致的、水墨淡彩的中国画。

白鹤的越冬地和繁殖地相当集中，世界上98%的白鹤都在鄱阳湖越冬，绝大部分都在俄罗斯西伯利亚的雅库特低地苔原上繁殖。

No sooner has the thick fog over the lake surface vanished in the early morning, the excited snow cranes (*Grus leucogeranus*) are already starting their singing and dancing show.With all other views in the background blocked by the misty air, the snow cranes are the only creature visible, presenting us with an elegant Chinese traditional ink painting.

The winter ground and breeding place of snow cranes are quite concentrated, with 98 percent of the snow cranes all over the world

spending winter in Poyang Lake and most of them
breed on the Yakut lowland tundra in Siberia,
Russia.

高原精灵·云南大山包
Plateau elf • Dashanbao, Yunnan

　　尽管这里是避寒的越冬地，云贵高原的大山包依然是冰雪的世界。黑颈鹤夫妇带着初次从繁殖地飞来的雏鸟顺利到达这里，兴奋得一起引吭高歌，只有旁边的孩子一头雾水，尚不能理解父母这种激动的心情。

　　黑颈鹤是世界上唯一生长、繁殖都在高原的鹤，它们在南边的云贵高原上越冬，在北边的青藏高原若尔盖等地繁殖，因此被称为"高原鹤"。在全球的15种鹤中，黑颈鹤是科学意义上发现得最晚的一种鹤。

　　Though it's a winter ground for birds to shelter from the cold winter, Dashanbao in Yunnan is still an ice-clad world. Black-necked crane couple make it here with their newly hatched baby from the breeding place and excitedly sing along the way, leaving their bewildered offspring wondering why their parents are in such a high mood.

　　The black-necked crane is the only crane that grows and breeds in the plateau. They spend winter in the Yunnan-Guizhou Plateau in the south and breed in Zoige on the Qinghai-Tibet Plateau, hence the name "plateau crane". Among the 15 species of cranes in the world, the black-necked crane is the latest to be found in scientific sense.

雪天的国宝•陕西洋县
National Treasure in the Snow • Yangxian County, Shaanxi

朱鹮有着鸟中"东方宝石"之称，洁白的羽毛、艳红的头冠和黑色的长嘴，加上细长的双脚，非常美丽。图为冬天雪后，朱鹮在河溪湿地里觅食。

1981年5月，中国科学家在陕西洋县发现7只朱鹮（1983年前，日本还有5只朱鹮，1983年后，中国之外再无朱鹮）。中国采取了强化就地保护为主、人工迁地保护为辅的正确策略，朱鹮种群数量得到了大大提高，脱离了濒危状态。

The Asian crested ibis (*Nipponia nippon*) is known as the "oriental jewel" among birds. Its white feathers, bright red crown, long black beak, and slender feet are very beautiful. The picture shows the Asian crested ibis foraging in river wetlands after a snowfall in winter.

In May 1981, Chinese scientists discovered seven Asian crested ibis in Yangxian County, Shaanxi Province (before 1983, there were five crested ibis in Japan, and after 1983, there were no crested ibis outside China). China has adopted the correct strategy of strengthening in situ protection, supplemented by artificial ex situ protection, and the population of Asian crested ibis has been greatly increased, and the bird is therefore lifted out endangered status.

雪中渔猎●吉林珲春
Fishing and Hunting in the Snow ●
Hunchun, Jilin

- -

　　大雪纷飞之际，白尾海雕抓住了一条鱼后正飞向空中，它会找一个高地或者在一根粗树枝上慢慢撕食这个战利品。

　　白尾海雕为大型猛禽，活动于湖泊、河流、海岸、岛屿及河口地区，主要以鱼为食，也捕食鸟类和中小型哺乳动物，常在水面低空飞行，即便是大雪纷飞也丝毫不影响它捕食的兴致。

The white-tailed sea-eagle (*Haliaeetus albicilla*) catches a fish and is flying to the sky. It will feast leisurely on its catch on a highland or on the thick branch.

White-tailed sea-eagles are birds of prey with large size that live in lakes, rivers, coasts, islands and estuaries. They feed on fish, as well as preying on birds and other small and medium mammals. They often fly low above the water and even heavy snowfalls won't dampen their strong appetites for foraging.

2 湿地其他野生动物
Other Wetland Wildlife

　　除了鸟类以外，湿地生态系统中的其他野生动物同样是相当丰富的，有兽类、两栖爬行类、鱼类等。

In addition to birds, a huge diversity of other wildlife species also live within wetland ecosystems, such as beasts, amphibians, reptiles, fish and so on.

水清鱼稠·贵州黔南
Clean Water and Abundant Fish •
Qiannan, Guizhou

　　在西南山地的山涧水溪中，由于山上植被茂密，其水质尤为清澈，鱼类自由自在地生活在这里。这种美丽的湿地景观，颠覆了"水清则无鱼"的俗话。

　　图中为云南光唇鱼，分布于珠江水系、长江中上游及其支流，常栖息于多石块的缓流水环境。杂食性，以丝状藻为主，水草次之。

Water of the streams and rivers flowing among the mountainous regions in southwestern China is particularly crystal clear due to the dense vegetation of the mountains. This kind of beautiful wetland landscape where fish abounds in the lucid water forcefully falsifies an old saying that claims "over-clean water breeds no fish".

The picture shows *Acrossocheilus yunnanensis* which typically lives in the Zhujiang River system, the middle and upper reaches of the Yangtze River and its tributaries. They are often found in niches characterized by rock-strewn slow-flowing rivers. An omnivores fish, they feed primarily on filamentous algae and other water plants.

水中嬉戏·四川唐家河
Playing in Water • Tangjia River, Sichuan

　　在唐家河清澈的河溪里，两只水獭在打闹嬉戏。它们忽而钻进水中，忽而跳上石头，这是活生生的水中精灵——在大白天碰上它们可不是一件容易的事。

　　水獭是哺乳动物，多穴居，白天休息，夜间出来活动，善于游泳和潜水，听觉、视觉、嗅觉都很敏锐，食性较杂，主要栖息于河流和湖泊一带，也有的生活在海岸、海岛水边。水獭是优良水环境的指示种。

　　Two otters (*Lutra lutra*) are romping and playing with each other in the clean streams of the Tangjia River. They are lively water elves, sometimes diving into the water and sometimes jumping onto the rocks. It is not easy to run into them during daytime.

　　Otters are omnivores mammals that mostly live in caves, sleeping during the daytime and coming out only in nights. As excellent swimmers and divers,they have extremely sensitive senses of hearing, vision and smelling. They mainly inhabit around rivers and lakes but occasionally are also found in coastal areas close to the sea and islands, making the man indicator species of high-quality water environment.

长江之神•安徽铜陵
The God of the Yangtze River • Tongling, Anhui

- -

长江江豚作为江豚的淡水亚种，已在地球上生存2500万年，目前主要分布在长江中下游及鄱阳湖一带，被称作"长江之神"和"水中大熊猫"。唐代文学家韩愈有诗云"江豚时出戏，惊波忽荡漾。"

长江是中国重要的生态宝库，是世界上水生生物多样性最为丰富的河流之一。这里，国家一级保护野生动物有长江江豚、白鲟、中华鲟、长江鲟、扬子鳄等，国家二级保护的野生动物有水獭、大鲵、胭脂鱼、秦岭细麟鲑等。

Yangtze finless porpoise (*Neophocaena asiaeorientalis*), a freshwater subspecies of finless porpoise known as "the god of Yangtze River" and "giant panda in the water", has been living on the earth for over 25 million years and is mainly distributed in the middle and lower reaches of the Yangtze River and Poyang Lake. Han Yu, a literary scholar of the Tang Dynasty, once wrote a poem, "Finless porpoise plays leisurely among rippling waves".

The Yangtze River is an important ecological treasure house in China and is one of the rivers with the richest aquatic biodiversity in the world. There are national first-class protected wild animals including Yangtze finless porpoises, white sturgeons (*Psephurus gladius*), Chinese sturgeons (*Acipenser sinensis*), Yangtze sturgeons (*Acipenser dabryanus*) and Yangtze crocodiles, as well as national second-class protected wild animals, such as otters (*Lutra lutra*), giant salamanders (*Andrias davidianus*), catostomidaes (*Myxocyprinus asiaticus*) and Qinlingtiny-scaled salmons (*Brachymystaxlenok tsinlingensis*).

妇唱夫随·广东湛江
A Loving Couple · Zhanjiang, Guangdong

在红树林海滩上，有一对鲎夫妇正在遛弯儿，和人们一般的判断相反，个大雄壮的是雌鲎，瘦小的是雄鲎。雌雄一旦结为夫妻，便形影不离，雄壮的雌鲎常驮着瘦小的丈夫蹒跚而行，享"海底鸳鸯"之美称。

鲎的祖先出现在恐龙尚未崛起的古生代，随着时间的推移，与它同时代的动物或进化、或灭绝，而只有鲎从4亿多年前问世至今仍保留其原始而古老的相貌，所以鲎有"活化石"之称。

On the mangrove beach, there is a couple of horseshoe crabs walking around. Contrary to popular belief, female horseshoe crabs are big and strong while male horseshoe crabs are the smaller ones. Once a male and a female get married, they become inseparable, with the stronger female horseshoe crab often hobbling along with her little husband. Therefore, they enjoy the reputation of "the lovebirds in the sea".

The ancestors of horseshoe crabs appeared in the Paleozoic Era that predates the rise of the dinosaurs. As time went on, their contemporaries either evolved into more advanced species or became extinct, but they still retained their primitive and ancient appearance they had ever since they came into being over 400 million years ago, which is why they are known as the "living fossils".

游"龙"•安徽宣城
The Wandering "Dragon" • Xuancheng, Anhui

扬子鳄是主要生活在湿地里的爬行动物，是中国特有的一种鳄鱼，也是世界上较小的鳄鱼品种之一。因其生活在长江流域，故称"扬子鳄"。在扬子鳄身上，至今还可以找到远古恐龙类爬行动物的许多特征。所以，扬子鳄又被称为"活化石"。

有研究证明，中华龙传说的原型来源于农耕文化江南地区的扬子鳄。

Yangtze crocodile, a reptile that lives mainly in wetlands, is a kind of crocodile unique to China and one of the smaller crocodile species in the world. As it lives in the Yangtze River Basin, it is called "Yangtze crocodile". Many features of ancient dinosaur-like reptiles can still be found on it, which earns them the honorable title of being the "living fossil".

Some studies have proved that the legendary archetype of the Chinese dragon is originated from the Yangtze crocodile of the farming culture in the Jiangnan area.

集体阳光浴·云南保山
Enjoying Sunbathing Together · Baoshan,
Yunnan

在高黎贡山自然保护区的山中小池塘里，一大群蝌蚪聚集到了水温较高的浅水区域，享受着春日的阳光。蝌蚪长得一点都不像妈妈，体呈圆形或椭圆形，没有四肢，而有一根小尾巴。

两栖类动物生活和繁育都离不开水，蝌蚪是其个体发育的初级阶段，生育的蝌蚪要在水中经过变态才能发育为成体的过程，被称为变态发育。

In a small pond in the mountains of Gaoligong Mountain Nature Reserve, a large group of tadpoles have gathered in the shallow area with warmer water to enjoy the spring sunshine. The tadpoles, which are of round or oval bodies that bear no limbs but a small tail, look anything but resembling their mothers.

Amphibians can not live and breed without water, and tadpoles are the primary stage of their individual growth. The process for tadpoles to transform from an immature form to an adult one is called metamorphosis.

不畏强暴·香港米埔
Standing Squarely up against the Bully · Mai Po, Hong Kong

一条弹涂鱼在海滩上蹦来跳去，这一下居然挡住了螃蟹的去路。小小螃蟹面对弹涂鱼的庞大身躯和胡作非为非常不满，直瞪着眼睛，不畏强暴地冲上前去。

退潮时的海滩往往是异常热闹的，无数的弹涂鱼、招潮蟹、小螃蟹、小贝类等在泥沼中忙碌着。体长不到成年人一指长的弹涂鱼，常依靠发达的胸肌匍匐或跳跃于退潮的滩涂上。

它除了用鳃呼吸外，还可以凭借皮肤和口腔黏膜来摄取空气中的氧气。

A mudskipper (*Periophthalmus cantonensis*) is jumping around on the beach, blocking the path of a tiny crab inadvertently. The offended little creature stares angrily at the massive body of the mudskipper and rushes boldly up to it, as if feeling angry about the unexpected and unjustifiable intrusion of the latter.

The beach at low tides is always full of jollification,with countless mudskippers, fiddler crabs (*Uca*), and small crabs and shellfish bustling in the mire. The mudskipper, whose body is less than the length of an adult finger, often creep or jump on the mudflats at low tide with its pectoral muscles. In addition to gill breathing, it can also rely on the skin and oral mucosa to take in oxygen from the air.

下海·山东长岛
Going to the Sea • Changdao, Shandong

清晨的阳光照在岛礁上，在岩石上休息了一整夜的西太平洋斑海豹，在早起海鸥的关注下，懒洋洋地下水了。

西太平洋斑海豹是唯一能在中国海域繁殖的鳍足类动物，尤为珍贵，有洄游的习性。它一生的大部分时间是在海水中度过的，仅在生殖、哺乳、休息和换毛时才爬到岸上或冰块上。

As the morning sun shines on the island reef, a spotted seal (*Phoca largha*), which has been resting on the rocks all night, lazily goes into the water under the attention of seagulls.

Spotted seals are the only fin-footed animal that can breed in Chinese waters and are particularly valuable. They are migratory animals that spend most of their life in seawater and climb to shore or on ice only at the time of reproducing, nursing, resting and changing fur.

远方在呼唤·海南西沙
Callings from Faraway Places • Xisha,
Hainan

　　雌性玳瑁白昼在海岸沙滩挖穴产卵，产完卵后它将回归大海，而雌性海龟产卵只在夜晚。

　　玳瑁主要的生活区是浅水礁湖和珊瑚礁区，珊瑚礁中的洞穴和深谷给它提供休息的地方。玳瑁主要捕食鱼类、虾、蟹和软体动物，也吃海藻、海绵类动物等。

　　Female hawksbill turtles (*Eretmochelys imbricata*) dig burrows to lay eggs on the coastal beaches during the day and return to the sea after fulfilling their duties, while other sea turtle species lay eggs only at night.

　　Shallow lagoons and coral reefs, which provides caves and deep valleys as resting places, are the main living areas for hawksbill turtles. They mainly prey on fish, shrimp, crabs and mollusks, but also eat algae and sponges.

3 湿地野生植物
Wild Flora in Wetlands

　　湿地的野生植物种类繁多，以一年生和多年生的草本植物为主。根据植物与水体之间的关系，可以直观地将它们划分为挺水植物、浮水植物和沉水植物。还有一些多年生的木本植物也生活在湿地里或者湿地边，这是我们讨论湿地植物时不能忘记的。

There are a wide variety of wild plants in wetlands, which are mainly annual and perennial herbaceous herbs. Based on their relationship to the water body, they can be visually classified as emergent plants, floating plants and submerged plants. There are also perennial ligneous plants that live in or along wetlands, which we must also take into consideration in discussions concerning wetland plants.

画意芦苇•辽宁双台子河口
Reeds as Beautiful as Paintings •
Shuangtaizi Estuary, Liaoning

- -

　　芦苇大概是湿地植物中分布最广、个子最高、数量和生物量都最大的挺水植物了。干净画面中一排芦苇及其清晰的倒影，确有中国水墨画的感觉。

　　芦苇、荻都是禾本科植物，是湿地里分布最广、最多的不同物种，有"北苇南荻"之说。杜甫的《秋兴八首》里那句"请看石上藤萝月，已映洲前芦荻花"，说明芦和荻远望是非常相像的两种植物。

Reeds (*Phragmites australis*) are probably the most widespread, the tallest, the densest-growing and the most biologically abundant emergent plants in wetlands. A row of reeds with their images clearly reflected in the water look just like a Chinese ink painting.

Both reeds and silvergrass (*Miscanthus saccariflorus*) are the most widespread and diverse grasses in the wetlands, just like the old saying "there are abundant reeds in the north and silvergrass in the south". Du Fu once wrote a sentence in his work Eight Poems for Autumnal Revival that goes roughly as follows: "Look at the moon above the vine on the rock, reed and silvergrass flowers on this island are reflected in the front", indicating that reeds and silvergrass are two plants that look very similar from a distance.

荻花遍地•山东黄河三角洲
The Ubiquitous Silvergrass Flowers • The Yellow River Delta, Shandong

----

　　人们常说芦荻，是把芦苇和荻混在一起来说的。其实，芦苇和荻分别是禾本科芦苇属和荻属不同属的植物，它们大多都长在湖泽、滩涂水岸边，都能长到2~3米高，茎叶也都是长条形，外形很相似。

　　我们可以从花序上看出明显不同，芦苇的花是圆锥形的，荻的花排列成指状。

People sometimes mistake reeds for silvergrass (*Miscanthus saccariflorus*), given that the two of them both grow along the shores of lakes and mudflats and have similar appearance, both capable of growing to 2~3 meters in height, both with long stems and leaves. But in fact, they are two different genera of plants in the grass family that belong respectively to Phragmites and Miscanthus.

In terms of the inflorescence, we can also see

a clear difference. The flowers of reeds are conical
while those of silvergrass are arranged in a finger
shape.

菱角与荇菜·江苏无锡
Water Caltrop and Yellow Floating Heart • Wuxi, Jiangsu

- -

　　水面漂浮的叶片，菱形形状的是一年生的水生植物菱角，形似睡莲的是荇菜，为多年生水生植物。

　　它们都是浮水植物，生于池沼、湖泊、沟渠、稻田、河流或河口的平稳水域，嫩茎都可食用，且菱角果肉也可食，在江南生长非常普遍。

　　The water caltrop (*Trapa natans*) is an annual water plant with rhomboidal leaves that float on water surface, while the yellow floating heart (*Nymphoides peltata*), shaped like a water lily, is a perennial aquatic plant.

　　They are both floating plants that grow typically in still or slow-flowing waters like ponds, lakes, ditches, rice fields, rivers or estuaries in southern China. Their tender stems of both them are edible, as is the pulps of water caltrops.

水下莛草·黑龙江兴凯湖
Underwater Pondweed • Xingkai Lake,
Heilongjiang

　　莛草为多年生沉水植物，生于池塘、湖泊、溪流中，在静水池塘或沟渠较多，与水中鱼儿相伴相生。

　　湿地中生长有大量的挺水植物（如芦苇）和浮水植物（如荇菜），都是比较容易被人们观察到的。但沉水植物只生长在水中，就近仔细看才能发现。

　　The curly-leaf pondweed (*Potamogeton crispus*), a perennial submerged plant which lives with fish in a symbiotic relationship, grows mainly in ponds, lakes, and streams, and is found more commonly in still ponds or ditches.

　　Wetlands are home to many emergent plants (such as reeds) and floating plants (such as yellow floating hearts), which are all easily observable to people. However, submerged plants only live under water and can only be found when you look carefully from close-up distance.

水葱·四川九寨沟
Soft-stem Bulrush • Jiuzhaigou, Sichuan

　　一片水葱出水犹如无数个正在跳舞的舞娘，若远看还似一片跳动的火焰燃烧在水面上。

　　水葱秆高大，圆柱状，最上面一个叶鞘具叶片。水葱花苞为秆的延长，直立，钻状，常短于花序。水葱主要生长在湖边或浅水塘中，为挺水植物。

　　Above the water surface, a group of soft-stem bulrushes (*Schoenoplectus tabernaemontani*) are dancing ballet, looking like a dancing flame burning on the water when viewed from distance.

　　Soft-stem bulrush is an emergent plant which mainly grows in lakeside or shallow ponds. Its culm is tall and cylindrical, with a leaf sheath at the very top. The bud is an extension of the culm, which is erect, subulate, and often shorter than the inflorescence.

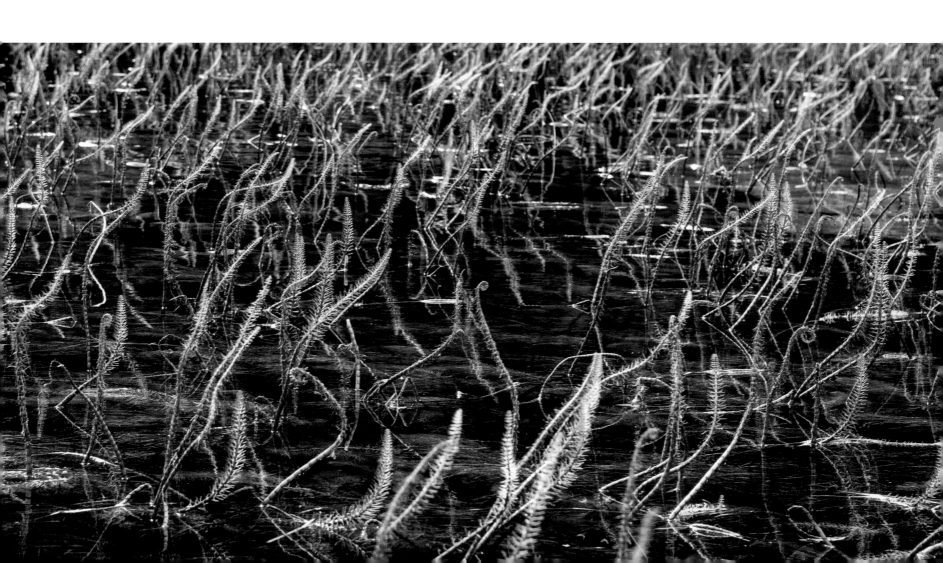

东滩的草·上海崇明东滩
Grass in Dongtan • Chongming Dongtan, Shanghai

崇明东滩湿地的先锋物种主要是海三棱藨草，随着土地淤积、高程升高，会逐渐被芦苇和互花米草所替代，但互花米草为外来物种。

崇明岛是中国第三大岛，是长江口规模最大、发育最完善的河口型潮汐滩涂湿地，在崇明岛最东端建有国家级自然保护区，以保护河口湿地生态系统和迁徙鸟类为主要保护对象的，2002年被列为国际重要湿地。

The pioneer species of Chongming Dongtan wetland is *Scirpus mariqueter*. But as the land elevates with the gradual sedimentation of silts, it will be replaced slowly by reeds and smooth cordgrass (*Spartina alterniflora*), the latter among which is an exotic species.

Chongming Island, the third largest island in China, is the largest and most well-developed estuarine tidal mudflat wetland in the estuary of the Yangtze River. A national nature reserve has been set up there for the protection of the estuarine wetland ecosystem and migratory bird sat the easternmost tip of theislandand was listed as an internationally important wetland in 2002.

沼泽草甸•黑龙江三江平原
Swamp Meadow • Sanjiang Plain, Heilongjiang

三江平原是我国湿地面积最大、类型最齐全的地区之一，其中，沼泽化草甸是主要植被类型，沼泽化草甸植物以小叶章最为普遍。小叶章草甸中是一片原始状态下的莲花塘。莲为多年生水生草本植物。

小叶章为多年生根茎型、寒温性禾草，对于气候变化、水文变化和人为活动极为敏感，是三江平原最有代表性的湿地草本植物。

The Sanjiang Plain is one of the largest wetlands with the most complete wetland types in China. Swamp meadow is the main vegetation type here, of which *Calamagrostis angustifolia* meadow is the most common kind to find. In the middle of the meadow is a primitive lotus pond overgrown withlotus (*Nelumbo nucifera*), a perennial aquatic herb.

Calamagrostis angustifolia is a perennial

rhizomatous that is the most representative wetland herb in the Sanjiang Plain. It prefers to grow in acold and temperate environment, and is extremely sensitive to changes in climate and hydrological conditions as well as to anthropogenic activities.

雨中水杉·广东广州
Metasequoia in the Rain • Guangzhou, Guangdong

- -

水杉是湿地中或湿地周边生长的木本植物，喜湿润，生长快，有"活化石"之称，对于古植物、古气候、古地理和地质学，以及裸子植物系统发育的研究均有重要的意义。

水杉是世界上珍稀的孑遗植物，过去科学界认为其早已灭绝，1941年因中国植物学家在四川万县谋道溪（今称万州区磨刀溪）首次发现而闻名中外。

The metasequoia (*Metasequoia glyptostroboides*) is a woody plant growing in or around wetlands. It likes moist environment and grows fast. Known as the "living fossil", it is important for the study of paleobotany, paleoclimate, paleogeography, geology and the phylogeny of gymnosperms.

The metasequoia, which was once deemed by scientists to have gone extinct, is a rare relict plant in the world. It was first discovered by Chinese botanists in 1941 in Moudao Creek of Wan County (now called Modao Creek of Wanzhou District), Sichuan Province and soon known throughout the world.

中国最大的红树 • 海南八门湾
The Largest Mangrove Tree in China • Bamen Bay, Hainan

据介绍，中国最大的红树就是图中这棵树。该树位于海南文昌的八门湾红树林自然保护区。这是退潮时红树林内部的状况。

八门湾红树林自然保护区有大约全世界40%的红树品种，是我国红树品种最多的地方之一。

According to the introduction, the mangrove tree shown in the picture is the largest one in China. It grows in the Bamen Bay Mangrove Nature Reserve in Wenchang, Hainan. Shown in the picture is the condition within the mangrove forest at low tide.

The Bamen Bay Mangrove Nature Reserve is home to over 40% of all the mangrove species found in the world, making it one of the places that are reputed for their richest diversity of mangrove in China.

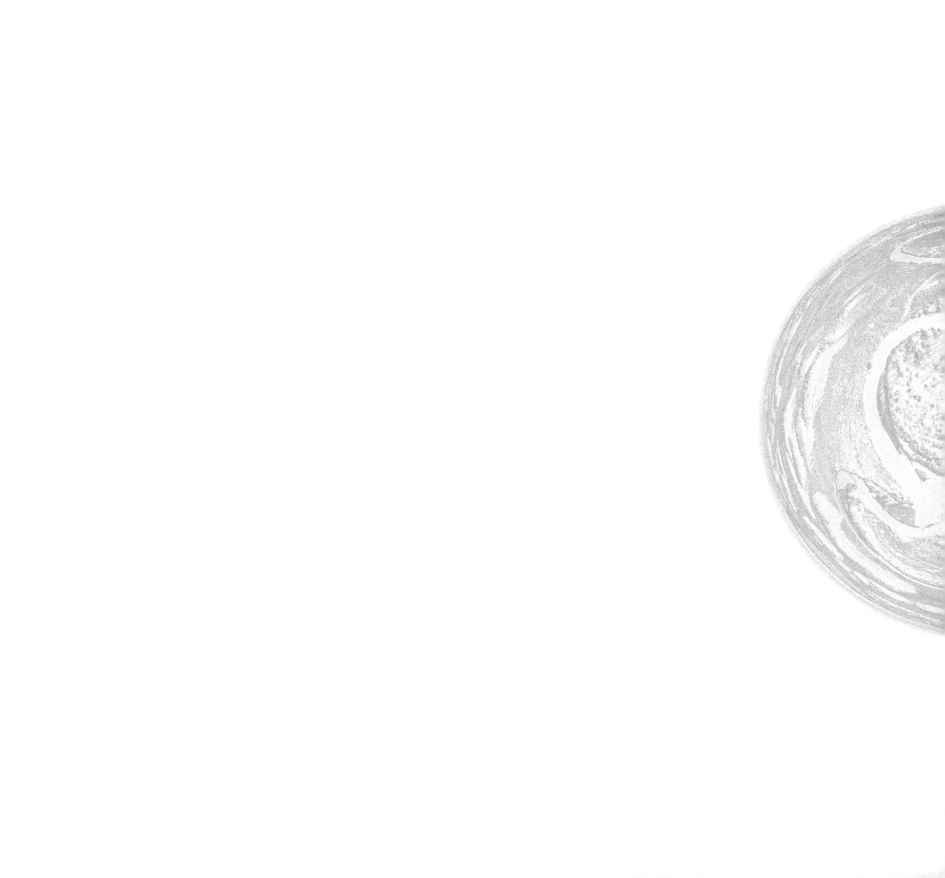

生命共同体

从整个地球自然环境来讲，水是连接所有生态系统的纽带，任何生态系统（森林、湿地、荒漠、草原、海洋等）都离不开水的滋润，这是对于"水是生命之源"的最好诠释。而湿地因水而生，无水而亡。由于水，湿地生态系统与地球其他生态系统都有着千丝万缕的联系，真正是"一损俱损，一荣俱荣"，生动地诠释了全球"生命共同体"的理念。这种特性，是其他生态系统都不可比肩的。

水和人类的关系更是相当密切，"逐水草而居"是人类生活的本性，而现今"绿水青山就是金山银山""山水林田湖草沙"系统协调的就是人类理想的"世外桃源"，我们现在正在进行的生态文明建设的目的就是要使"人与自然和谐共生"的愿景早日实现。

In terms of natural environment of the Earth, water links all the ecosystems and all ecosystems (forest, wetland, desert, grassland, sea, *etc.*) depend on the nourishment of water, which is the best demonstration of "water is the source of all lives". The wetland thrives because of the nourishment of water and dries up because of the lack of water. Because of water, wetland ecosystem and other ecosystems on the Earth are intertwined with each other. In fact, they all rise and fall together, which is a vivid manifestation of the concept of global "community of lives". No other ecosystems can compare with wetland ecosystem in this regard.

The bond between water and human beings is even tighter. It is human nature to live in places with abundant water resource and verdant grasses. Nowadays, the pursuit of "lucid waters and lush mountains are invaluable assets" and the system of "mountains, rivers, forests, farmlands, lakes, grasslands and deserts" emphasizes that human beings dream of a haven of peace. And the aim of our ongoing ecological civilization efforts is to bring the vision of "harmonious co-existence of man and nature" into reality in the earliest possible time.

COMMUNITY OF SHARED FUTURE FOR ALL LIVES

1 湿地与森林
Wetlands and Forests

　　森林是陆地生态系统的主体，无论海拔、纬度如何变化，无论光热水条件如何不同，无论是在干旱荒漠中还是热带雨林里，森林都是那个子最高、生物量最大之所在。生物量大，蒸发量就大；吸水量强，储水能力就强。因此，森林调节水的能力最强。森林与湿地的关系密切，它们之间有时正相关、有时负相关。

　　The forest is the main body of terrestrial ecosystem. No matter how altitude and latitude vary, no matter how different the condition of light, heat and water source is, and be it in deserts or rainforests, the forest is always the thing that stands the tallest and holds the largest biomass. Larger biomass can lead to higher volume of evaporation; and greater ability to absorb water means greater water storage capability. Consequentially, the forest is the most competent to conserve and regulate water resources. Forests and wetlands are closely tied to each other. Sometimes they are positively related and sometimes negatively.

湖岗森林•黑龙江兴凯湖
Forest over Lakeside Mound • Lake Khanka, Heilongjiang

这是处于兴凯湖大湖、小湖之间的湖岗，湖岗长30余千米，宽100~200米。两片湿地夹一片森林，既养育了森林，又让这片森林在相对隔离的特殊环境中生长，使湖岗上的森林演替以一种独特的方式进行，形成了特有的树种——兴凯赤松。

兴凯赤松喜生于排水良好的湖岗沙地，具有抗旱、抗风、抗寒和耐土壤瘠薄等特性，能耐零下40摄氏度低温。

Situated between the big and the small Khanka Lakes, this lakeside mound measures over 30 kilometers in length and 100 to 200 meters in breadth. Tucked safely between the two wetlands, this forest here is not only self-sustaining, but also brings into existence a relatively isolated special niche for it to thrive and flourish. Therefore, the forest by this lakeside mound evolves in a special way, contributing to the birth of *Pinus takahasii*, a tree species unique to this place.

The sandy land of the lakeside mound provides a favorite site for *Pinus takahasii*, which is characterized by its higher resistance against drought, strong wind, barren soil and down-to-minus-40-degree low temperature.

林中沼泽·四川王朗
Marshes amidst Forests • Wanglang National Nature Reserve, Sichuan

----

图中显示沼泽中有倒伏并已经腐烂的粗大树干，这说明原来这片森林是相当茂盛的。

如果森林水分下渗困难，地表过湿，容易引起林地沼泽化。林下沼泽或林间空地的沼泽不断向四周扩展，恶化了树木的生长环境，造成树木大量死亡，就会形成森林沼泽。这时森林和湿地之间为负相关。

From this picture, rotten and strong trunks can be seen fall in the marshes, indicative of robust growing condition of this forest in the past.

If moisture in forest has difficulty sinking into deeper underground and the surface of ground becomes too wet, then forest land will be prone to turn into swamps. When understorey swamps existing among forests or over glades spread, the living environment of trees will deteriorate and they will perish in large number. In the end, the forest marshes will form. This is a good case where forest and wetland are negatively related.

晚秋的画 • 新疆塔里木
A Painting of Late Autumn View • Tarim, Xinjiang

晚秋的胡杨林重重地染上了深黄色，森林与湿地相互依偎，倒映在平静的塔里木河上的影像显得如此的协调，这是在极干旱的南疆荒漠地区难得一见的景观。

胡杨林是中型的落叶阔叶乔木，树高能达10~15米，直径可达1.5米，耐旱耐涝，生命顽强，是生长在极干旱地区少有的乔木树种。此时，森林与湿地之间为正相关。

In late autumn, *Populus euphratica* forest is tinted with dark yellow. Forest and wetland cling intimately onto each other. The reflections of the towering trees in the Tarim River fit so naturally and compatibly into the surrounding. This is indeed a rare view to find in the extremely dry desert areas in the southern part of Xinjiang.

Populus euphratica is a deciduous and broadleaf tree species of medium height. Its height amounts to 10 to 15 meters and diameter 1.5 meters. This tenacious tree species, rarely seen in extremely dry area, can withstand drought and flood. This is a good case where forest and wetland are positively related.

海岛植被·海南探航岛
Vegetation on the Island • Chenhang
Island, Hainan

地处南中国海中星星点点的小岛上，在热带阳光和季风降雨的呵护下，珊瑚砂构成的小岛上及潟湖湿地旁的森林植被郁郁葱葱。

岛上树种有厚藤、草海桐、棕榈和椰树等。

Over the numerous small islands sparsely scattered across the South China Sea, forests and vegetation on small islands made of coral sand and by the lagoon wetland flourish with vigor, thanks to the nourishment of tropical sunlight and monsoonal rainfall.

Tree species on this island includes bayhops (*Ipomoea pes-caprae*), beach cabbage (*Scaevola sericea*), palm (*Trachycarpus fortunei*), coconut tree (*Cocos nucifera*) and other species.

秋染月亮湾•新疆喀纳斯
The Moon Bay Tinted with Autumn
Colors • Kanas, Xinjiang

　　一层淡淡的薄雾悄悄地飘浮在月亮湾上，碧蓝的河水在秋色斑斓的森林里扭身、旋转而过，像清晨初见的哈萨克姑娘，在向我们展示着她对人生美好的憧憬和向往。

　　秋天的森林和流水告诉人们，这是一年中色彩最丰富、最美丽的季节。

　　A thin layer of mist floats on the Moon Bay quietly, and the blue river meanders in forest tinted with autumn color, as if you catch a glimpse of a Kazakh girl in the early morning, who is demonstrating her beautiful vision and aspiration for life to us.

　　It looks as if the forest and flowing water were reminding people: autumn is the most colorful and beautiful season of the year.

五花海秋 • 四川九寨沟
The Wuhua Lake in Autumn • Jiuzhaigou, Sichuan

　　俗话说"九寨归来不看水"，九寨沟里有大大小小一百多个湖泊，湖水清澈如镜，森林和湿地相映成趣、相互依托。

　　这里每个湖都有每个湖自己的特点，四季变幻无穷。湖水的变幻其实就是湿地植物生长的变幻，更是周边森林色彩的变幻。此时湖边森林秋色正浓，映衬着湖水更加妩媚动人。

　　As a popular folk saying goes that, "one will forget all the other lakes after visiting Jiuzhaigou." There are up to one hundred variously-sized lakes in Jiuzhaigou. The lakes are as limpid as mirrors. Together with forests, the wetlands form a beautiful picture.

　　Every lake has its distinctive feature and shows different appearance in different seasons. The change of appearance of the lakes is virtually the changes of the growing of wetland plants and the color of neighboring forest. Now the forest by the lakes is tinted with autumn colors, making the lakes more beautiful.

雾凇森林 • 吉林长白山
RimyForest • Changbai Mountains, Jilin

　　在我国东北地区冬天不封冻的湿地边，雾凇这个美丽的森林景观常常出现。

　　雾凇非冰非雪，是雾中无数零摄氏度以下而尚未凝华的水蒸气随风在树枝等物体上不断积聚冻粘的结果。雾凇形成需要气温很低、水汽又很充分，同时能具备这两个极重要而又相互矛盾的自然条件是难得的。

Rime, a beautiful forest scenery, can be frequently seen on wetlands that won't freeze in winter in northeast China.

Rime is neither ice nor snow. Instead, it is formed when the vapour in mist, whose temperature drops below 0°C but hasn't yet desublimate, clings tightly onto tree branches and other objects and gradually accumulates into clusters. The formation of rime depends on the presence of extremely low temperature and abundant moisture in the air. It is indeed a very rare occasion when both the two critical but mutually conflicting natural conditions are met simultaneously.

冬天林区里的河 • 黑龙江漠河
Rivers in Forest Areas in Winter • Mohe River, Heilongjiang

　　白雪给大地铺上了银装，所有物体都变成了简单的蓝白两色。平时蜿蜒流动在林区的河流已经冻结，把她滋润的这片森林的妙曼身姿毫无遮挡地呈现了出来。只有深深地热爱着这片土地，她才会不断转身回眸，欲走还留。

　　冬天是森林最不活跃的季节，雪和冰河仍然依偎着它。因为雪和冰河知道，一旦天气转暖，春天到来，森林最需要的就是它们的帮助。

　　Snowfall covers the earth up with a blanket of white, turning everything into a color of plain blue or white. The river that used to meander quietly through the forest is now frozen, which brings into sharp relief the fullest glory of the forest that it nurtures. She lingers on, reluctant to bid her farewell to the land, only because she loves this latter so dear.

　　Although forest is very inactive in winter, snow and icy rivers will still keep company with it, because they know what forest needs most is their help once temperature gets higher when spring comes back.

2 湿地与草原
Wetlands and Grasslands

草原是众多大江大河的发源地和水源涵养区，孕育了众多的湖泊沼泽湿地。在草原上，河流弯弯曲曲，湖泊星罗棋布，沼泽常生常伴。只要有土有水，草本植物就是这里最先长出来的优势物种，水多时是湿地，水少时是草原。在中国北方，当地人多用泡子、淖等来称呼那些浅浅的，时有时无、时大时小，变幻无常的草原湿地。

Grasslands are often the sources and water conservation areas of many major rivers, not to mention the numerous lakes, marshes and wetlands they nurture. There are twisting rivers and dotted lakes on grasslands, and marshes can be frequently seen there. As long as there are soil and water, herbaceous plants will be the first dominant species to come about. In times of abundant water supply, they appear in the form of wetlands; when water is scarce, it will take the shape of grassland. In northern China, these shallow grassland wetlands are often known among local people aspaozi, nao, denoting the frequently changing status of this ecosystem that turns up unexpectedly for a while and then vanishes all of a sudden, or expands and shrinks in size totally on its own course.

草原湿地•内蒙古苏尼特
Grassland Wetland • Sonid, Inner Mongolia

这是草原还是湿地？你说是草原，它就是草原；你说是湿地，它就是湿地。草原上的这类地方最符合《湿地公约》中"季节性的、暂时有水的地方"的概念。

夏季降雨，草原上低洼的地方就会积一层水，雨季过了就会慢慢消失。在这个过程中，积水面会有或大或小的变化。这片湿地的边界具体在哪里？很难说得清楚。在许多地方，草原和湿地就像一对孪生兄弟，相伴相生。

Is it grassland or wetland? The answer depends totally on how you prefer to name it. Call it grassland or wetland, and it can be whichever way you call. Places of this type over the vast grasslands fit neatly into the definitions stated in the *Convention on Wetlands*, which goes as follows: "places temporarily or seasonally filled with water".

With the arrival of abundant rainfall in summer, water ponds will appear in low-lying placed on grasslands, and when the rainy season is over, they will disappear with it. In this process, the layer of water will vary in coverage. So where's the specific boundary of this wetland? It's a question hard to answer. In many places, grasslands and wetlands are just like twin brothers that are constantly hanging around together, making it difficult for others to tell who is who.

肥美的当雄草原•西藏拉萨
The Lush Damxung Grassland • Lhasa, Tibet

　　当雄，藏语意为"心目中的草原"，是高原寒温带半干旱季风气候区里的草原，主要湿地有桑曲河、拉曲河和若干湖泊沼泽，这里是拉萨河主要水源供给区域，是难得的肥美草原。尽管牧草生长期短，仅90~120天，但这个期间是牛羊抓紧上膘储存营养的关键期。

　　西藏是我国草原大省（自治区），境内的草原中河流、湖泊、沼泽等湿地的面积占比很大。

Damxung, which in Tibetan language means "grassland in one's mind", is located in semi-arid monsoon climate region on plateaus of cold temperature zone. The major part of this wetland ecosystem consists of the Sangqu River, the Laqu River and many other lakes or marshes, which jointly make up the chief source of the Lhasa River as well as a lush pasture that is not often seen on the plateau. Although the growth period of herbage is short, lasting only for 90 to 120 days, this period is a crucial time span that cows and sheep can by no means lose in order to put on sufficient weight and store enough nutrition.

In China, Tibet is an autonomous region known for its vast grasslands. Rivers, lakes, marshes and other wetlands take up a large proportion of grasslands in Tibet.

Community of Shared Future for All Life

湿地牧场·新疆喀纳斯
Wetland Pasture • Kanas, Xinjiang

这画面是哈萨克牧民从阿勒泰山里的夏草场转到山下的冬草场，途经喀纳斯河时把牲畜放开吃草的场景。在如此风景秀丽的地方转场放牛羊，看着就不禁让人深深陶醉。

喀纳斯河从喀纳斯湖流出后，途经一长段地势比较平坦的河谷，河道两边产生了大片的草原、沼泽和森林。

Captured in the photo is the scene when Kazakh herdsmen are transferring from summer pasture on the Altai Mountains to winter pasture at the foot of the mountains, during which the livestock are left free to graze on the pasture lands as they pass by the Kanas River. In such a beautiful place, people can't help but be intoxicated to look at the cattle and sheep roaming care-freely over the grassland.

Following it departure from the Kanas Lake, the Kanas River flows through a long valley that is relatively flat, leading in result to the formation of numerous grasslands, marshes and forests on either bank of the river.

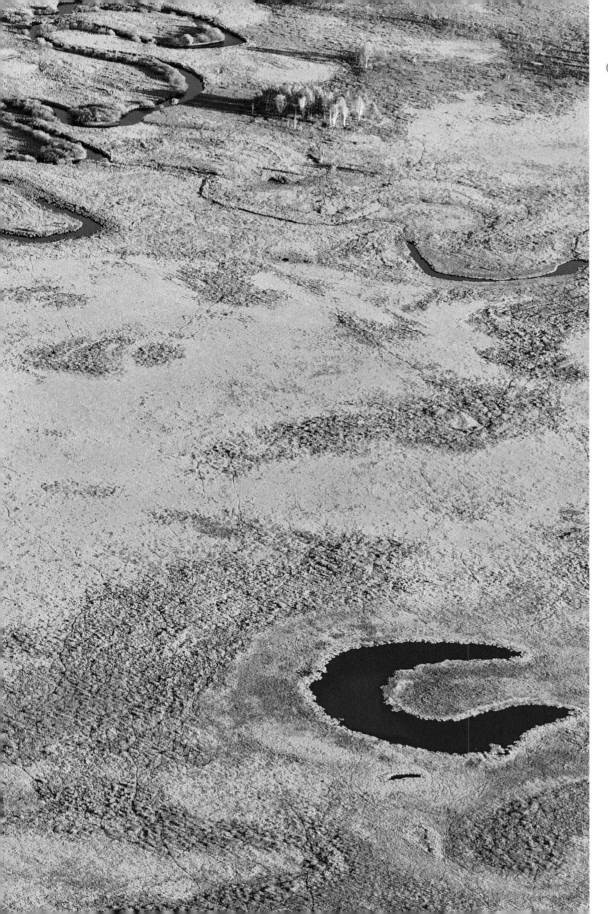

蜿蜒草原河•黑龙江南瓮河
Winding River on Grassland • Nanweng River, Heilongjiang

　　秋天的嫩江流域，由于地势平坦且有冻土层的存在，河流流速平缓、蜿蜒曲折。在如地毯般的金黄色草原上，河划出了一道复杂的蓝色曲线，来来回回、扭来扭去，陆被扭成了河，河被扭成了湖；扭出了岛状森林，扭出了牛轭湖泊，扭出了其他地方都难以看到的特殊景观。

　　东北草原区河流湖泊众多，是松花江、嫩江、额尔古纳河等河流重要的水源涵养区，仅呼伦贝尔境内就有大小河流3000多条，其中，流域面积在100平方千米以上的河流就有550条。

　　Due to the flat terrain and the existence of permafrost, the Nenjiang River meanders across the lands at a relatively slow pace in autumn. As it twists and turns through the blanket-like golden grassland, the river leaves behind it an elegant curve, occasionally in the form of river and occasionally in the form of lake, which jointly present viewers with a unique scenery that no other place can compare.

　　There are numerous lakes in the northeastern grassland area, which is an important water conservation area for the Songhuajiang River, the Nenjiang River, the Argun River and others. In Hulunbuir District alone, there are over 3,000 variously-sized rivers, 550 among which come with a catchment that extends for over 100 square kilometers.

乌拉盖草原·内蒙古锡林郭勒
Wulagai Grassland · Xilingol League,
Inner Mongolia

　　乌拉盖草原位于锡林郭勒大草原的东北部，是世界上保存最完好的天然草原之一，属森林草原向典型草原的过渡地带，是横跨欧亚大陆、延绵数千千米"典型草原"的东端。草原动植物种类繁多，轰动一时的"狼图腾"的故事就发生在这里，也是法国导演据此执导的电影《狼图腾》的重要取景地。

　　草原具有强大的水土保持功能，是保持水土的"先锋卫士"。据监测统计，草地比裸地的含水量高20%以上，在大雨状态下草原可减少地表径流量47%~60%，减少泥土冲刷量75%左右。

Situated on the northeastern part of Xilingol Steppe, the Wulagai Grassland is one of the most well-maintained natural grasslands around the world. Falling within the transitional zone between forest grasslands and typical grasslands, it makes up the east tip of the "typical grassland" that extends across the Eurasia continent for thousands of kilometers. In addition to being a haven for a rich variety of wild flora and fauna species, this is also the place where the world's famous wolf totem story took place, on basis of which a French director produced the blockbuster movie *Wolf Totem*.

An active player in soil and water conservation, grasslands are often hailed to be the "pioneer guard" in this aspect. Statistics generated from field monitoring indicates that grasslands, compared with their bare counterparts, are capable of conserving 20% more water, cutting down surface runoff by approximately 47%–60%, and sediment erosion by 75% when heavy rains pour down.

3 湿地与荒漠
Wetlands and Deserts

　　湿地与荒漠，一湿一干，在外表上就给人们一种毫不相干的印象。其实，在自然界的很多环境条件下，湿地与荒漠是相互依存、相互支撑的，它们之间的关系符合大自然对立统一规律。

　　Wetlands and deserts, one being wet and the other dry, often give people an apparent impression that they are mutually exclusive and have little to do with each other. But the fact is: wetlands and desert in nature are mutually dependent on many occasions, conforming perfectly to the natural law of unity of opposites.

新的鸬鹚岛·青海青海湖
New Cormorant Island • Qinghai Lake, Qinghai

乌云密布，不断前进的流动沙丘正在向青海湖腹地进发，倒是给挤在鸬鹚岛上的鸬鹚们拓展了大片新的栖息地。

一方面青海湖流动沙丘的扩展日益引起人们的注意；另一方面由于气候变暖，冰川融水加快，入湖水增加使湖面上升，青海湖面积越来越大了。这第二方面多少掩盖了第一方面的问题，近年来青海湖发生的这类现象值得我们警惕。

Under the dark clouds that are gathering in the sky, sandy dunes are increasingly encroaching upon the heartland of the Qinghai Lake. But nevertheless, as the saying goes, it is an ill wind that brings no one good, the progressively advancing sandy dunes is also expanding the habitats of cormorant (*Phalacrocorax*) that have long settled down on the Cormorant Island.

On the one hand, people are increasingly concerned about the encroachment of active sandy dunes onto the Qinghai Lake; on the other hand, as a consequence of the increasing in-flow caused by global warming and glacier melting, the water level of Qinghai Lake is also rising, accompanied by a notable expansion in its surface size. Given that the latter has more or less covered up the urgent-ness of the former, what is happening in the Qinghai Lake in recent years deserves our attentions.

沙水共存·内蒙古科尔沁沙地
Co-existence of Sand and Water · Korqin Sandy Land, Inner Mongolia

----

　　黄色的沙土和白色的冰冻水面相映成趣。冬日的科尔沁沙地上荒漠与湿地的相互关系更显紧密，整个画面透露着苍凉。

　　科尔沁沙地在大兴安岭以东的西辽河平原，比起大兴安岭以西的锡林郭勒草原，水分、光热条件都要丰富得多。历史上，这里曾是水草丰美、牛羊肥壮的科尔沁草原，由于气候变化及过垦过牧，生态环境严重失衡，变成了中国最大的沙地，是京津冀风沙的主要源头之一。

The yellow sand and the white frozen water form a delightful contrast. The tight bond and mutual dependence between deserst and wetlands are highlighted all the more notably by the bleak but awesome winter view of the Korqin Sandy Land.

The Korqin Sandy Land is located at the West Liaohe Plain that lies to the east of the Greater Khingan Mountains. Compared with the Xilingol Grassland that lies to the west of the Greater Khingan Mountains, the Sandy Land here enjoys more moisture, sunlight and heat. In ancient times, this used to be the site where the beautiful and lush Korqin Grassland, together with its thriving husbandry businesses, was found. However, due to the joint impacts of climate change, overgrazing and abusive land use, the ecological environment balance here has degraded so much as to turning it into the largest sandy land in China and the major source of sandstorm in Beijing-Tianjin-Hebei region.

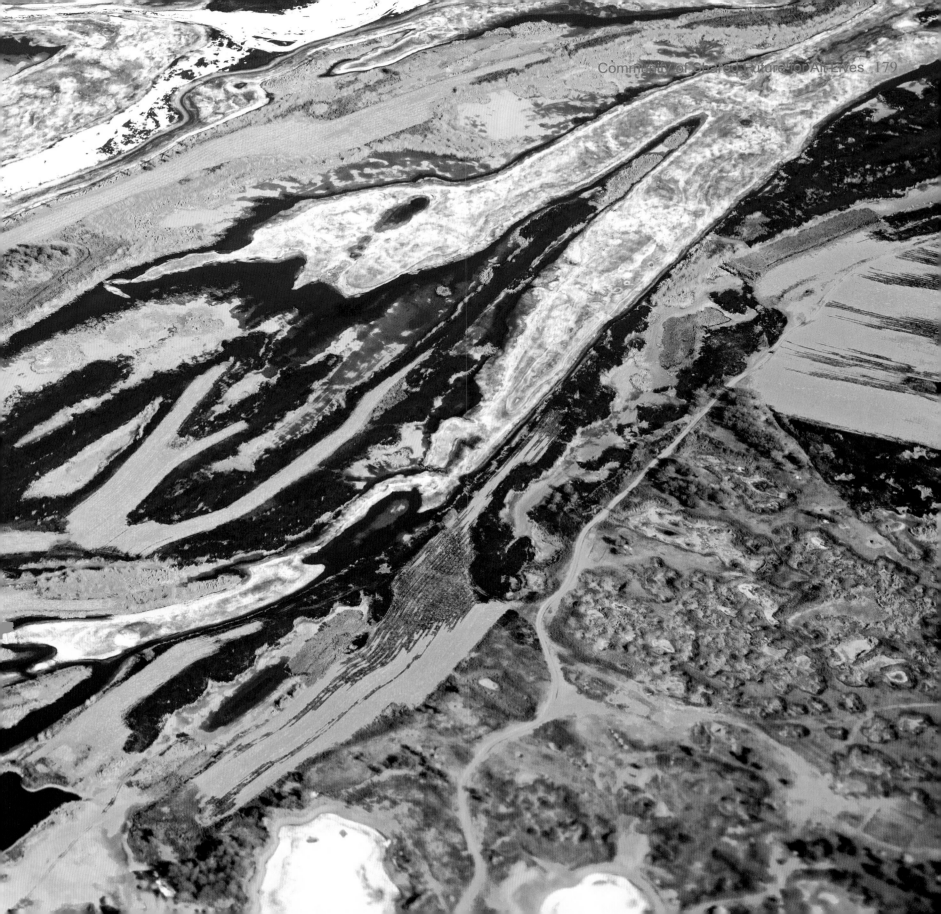

生态修复·内蒙古通辽
Ecological Restoration · Tongliao, Inner Mongolia

　　这里的沙地经过生态修复，很快就焕发出了勃勃生机，湿地周围草木兴旺、满眼葱茏。

　　在中国四大沙地中，科尔沁沙地的光热水条件最为优越。人们一边治理荒漠化，一边抓好湿地的修复是最为有效的综合治理办法。其实，科尔沁原本就是一片水草肥美的草原，现在人们所做的只不过是"师法自然"——按照自然规律去做，用人为的力量进行修复，尽量恢复到被破坏前的样子。

　　Thanks to the implementation of massive ecological restoration projects, the sandy land here is putting on a vibrant new look, with lush grasses and verdant trees steeping the wetland.

　　Of the four great sandy lands in China, the Korqin Sandy Land is blessed with the best condition of sunlight, heat and water. The comprehensive solution of combing desertification control efforts with wetland restoration is the most effective. Indeed, Korqin is essentially grassland with rich resources of water and grass. Therefore, what people do now is just learning from the nature—carrying out artificial restoration efforts while observing earnestly to the natural law and trying our best to restore it to what it was like.

沙漠与海子 • 内蒙古巴丹吉林沙漠
Desert and Lakes • Badain Jaran Desert, Inner Mongolia

　　巴丹吉林沙漠是中国八大沙漠之一，更由于这里湖泊星罗棋布，湖泊湿地中芦苇丛生，树木挺立，水鸟嬉戏，鱼翔浅底，而享有"漠北江南"之美誉。

　　干旱的沙漠与饱水的湿地看似是相互矛盾的两个方面，其实，高大的沙丘就是一个巨大的储水库，荒漠和湿地之间体现的是大自然的对立统一规律。

Badain Jaran Desert is one of the eight great deserts in China. It is also reputed as "a place in Mobei (north of desert and Gobi in northern China) that comes with the scenery of Jiangnan (area to the south of the Yangtze River)" for the dense distribution of lakes and wetlands where flourishing reed groves, verdant trees, carefree waterfowls and leisurely swimming fish can be readily seen.

Dry desert and moist wetland seem to be two mutually contradictory things. In fact, the tall dunes are actually an enormous water reservoir.

Thus, the relation between desert and wetlands is a demonstration of natural law of unity of opposites.

湖心盐壳地·新疆罗布泊
Salt Crust in the Center of the Lake • Lop Nur, Xinjiang

- -

沧海桑田，世界上湿地变荒漠规模最大、面积最广的地方应该是罗布泊了。罗布泊是蒙古语音译名，意为"多水汇集之湖"。古罗布泊湖面积曾经超过1万平方千米。图中所示现今的湖底不仅干涸，而且遍地都是坚硬且翘起的盐壳。

在卫星上看干涸的罗布泊，形状宛如人耳，是湖水步步干涸退后的痕迹形成的，因而罗布泊被誉为"地球之耳"；又因为这里是极干旱气候，寸草不生，人迹罕至，也被称作"死亡之海"。

Lop Nur may be the one place in the whole world that has undergone a fundamental transformation from being a resilient and far-stretching wetland to a massive desert. Lop Nur is derived from a word in Mongolian that literally means "a lake where several rivers merge". In

ancient times, the coverage of Lop Nur used to exceed 10,000 square kilometers. It can be seen from the picture that the Lop Nur of the present has dried up into a barren land strewn with hard and upturned salt crusts.

Looked from satellite, the dried-up Lop Nur is like the ear of a person. This is formed with the trace of the lake's retreating and becoming dry. Thus, Lop Nur is known as "the Ear of the Earth". Extremely dry, plant-less, and almost inaccessible to human beings, it is also known as "the Sea of Death".

沙漠天鹅湖 • 内蒙古阿拉善左旗
Swan Lake in Desert • Alxa Left Banner, Inner Mongolia

金色连绵的沙丘和绿色的榆树林倒映在清澈的湖面上，湖水中大天鹅在翩翩起舞，多么宁静、祥和而又充满生气的美丽画面啊！

在干旱、半干旱的中国西北沙漠中，沙漠并非是没有生气、活力之地。荒漠是地球上客观存在的一种生态系统，荒漠中的湿地和野生动植物是荒漠生态系统中不可或缺的有机组成部分。

Against the backdrop of the rolling golden dunes that extend until the horizon, a crystal-clear

lake is tucked gently among a verdant forest of elm trees (*Ulmus pumila*), in which groups of whooper swans (*Cygnus cygnus*) are dancing elegantly. What a stunning and peaceful view steaming with the vibrancy of life!

The deserts in the dry and semi-arid desert in northwestern China are by no means stagnant

and lifeless. As a unique ecosystem of the Earth, wetlands and wildlife make up an integral part of the desert ecosystem that is indispensable for the health and wellbeing of the planet

4 湿地与海洋
Wetlands and Seas

　　低潮时水深不超过6米的海域就是滨海湿地，滨海湿地其实就是大陆和海洋的交汇地带，包括沿海滩涂、红树林地等。湿地与海洋的交汇是相互交融、无法分割的。

　　Coastal wetlands refer to sea areas where its water depth is less than or equal to 6 meters at low tide. It is actually the convergence zone between land and ocean, including beaches near the ocean and areas with mangroves. It is impossible to draw a clearly-defined line between wetlands and seas.

黄渤海分界线·山东长岛
Boundary between the Yellow Sea and the Bohai Sea • Changdao Island, Shandong

----

在南长山岛的长山尾处，人们能够清楚地看到我国渤海和黄海的分界线。一道沙洲将碧海分为两个世界，来自黄海和渤海两个方向的海水在这里相遇、相撞，激起道道浪花，令人无不感叹这大自然的神奇造化。

庙岛群岛是散布在胶东半岛和辽东半岛之间的一系列岛屿，岛与岛之间的连线是我国渤海和黄海的地理分界线，南长山岛是庙岛群岛中最大的岛屿。

Standing at the end of the South Changshan Islands located in the southern part of Changshan County, people could see clearly the boundary between the Yellow Sea and the Bohai Sea, where a sandbank divides the ocean into two parts. Seawater derived from these two rivers meets and collides, stirring up spoondrift. Everyone who has seen it marvels at the magical power of nature.

The Miaodao Islands are a string of islands which are located between the Jiaodong Peninsula and the Liaodong Peninsula. The geological dividing line between the Yellow Sea and the Bohai Sea is the channel between the two peninsulas. The South Changshan Islands is the largest islands among the Miaodao Islands.

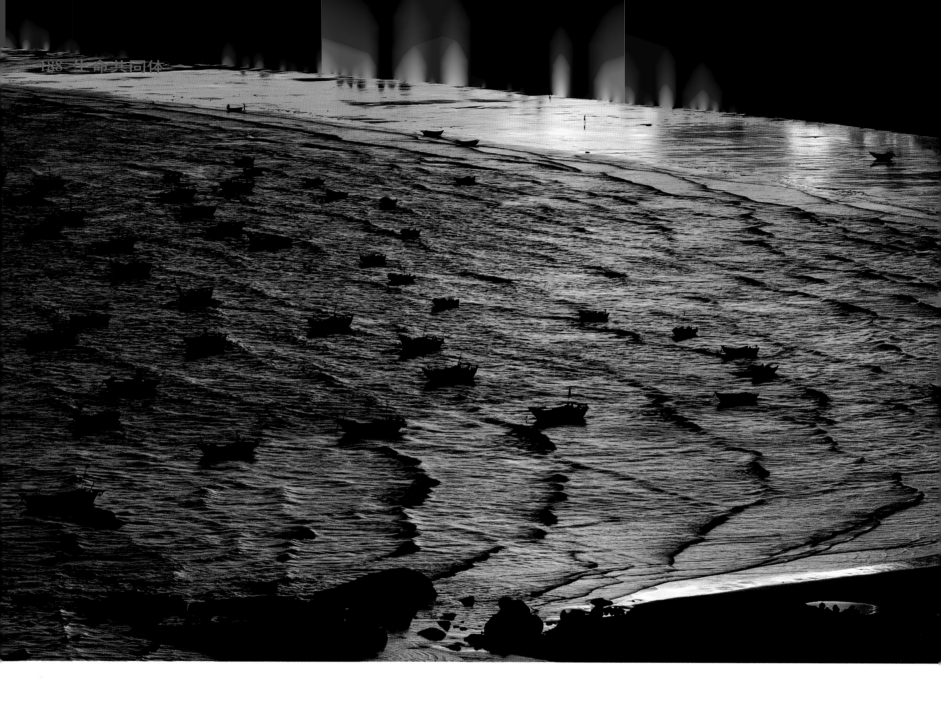

金色港湾•广东湛江
The Golden Harbour • Zhanjiang, Guangdong

涨潮了，旭阳下的湛江沃内海海湾风浪阵阵，海水泛着道道金光。渔船已经回港定锚，任凭海浪不断地晃动着它们，赶海的人也都陆续回家，渔港繁忙的一天就要结束了。

人们根据大海潮涨潮落的规律，赶在潮落的时机到海岸的滩涂和礁石上打捞或采集海产品的过程被称为赶海。每一次潮起潮落期间，都有无数大海留在海滩上的馈赠。

When tide rises, the inner sea's bay in Zhanjiang under the sunrise is blowing with wind and waves, and the surface of sea glowing with golden light. The fishing boats have already returned to the port and anchor, swaying in rhyme with the waves that tap them gently. Sea-goers have also gone home in succession, and the busy day in the fishing port is drawing to a close.

Sea-going refers to the activity when people,

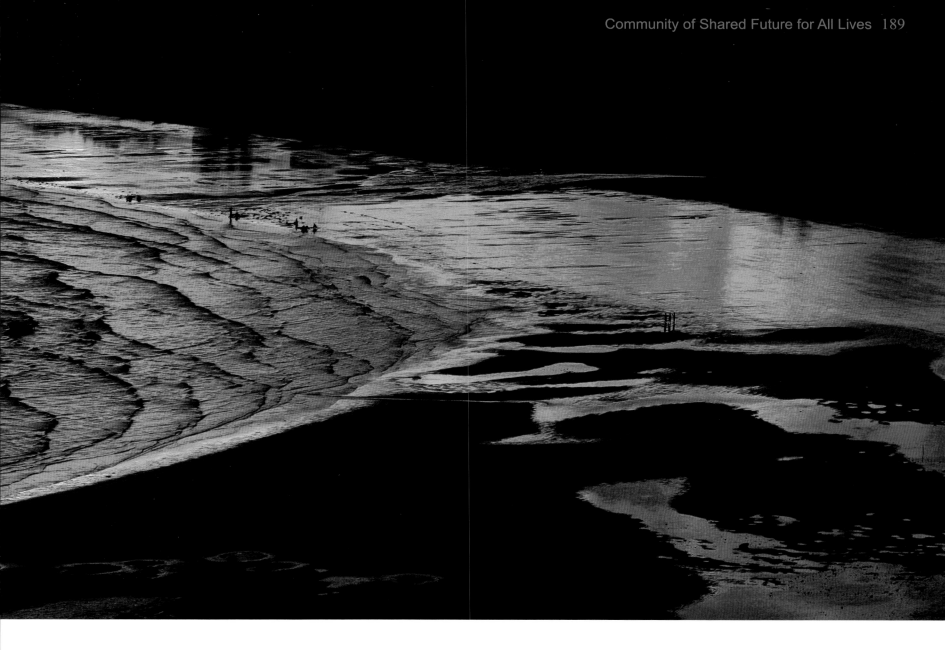

through watching the ebb and flow of the sea, take advantage of the time window when tides are at their ebb to salvage or collect seafood from the beaches and reefs on the coast. During every ebb and flow, there are countless gifts that the sea leaves on the beach.

鸟瞰西沙洲·海南西沙洲岛
A Bird-eye View of the West Sand Islet ·
the West Sand Islands, Hainan

　　西沙洲岛由珊瑚堆积而成，树木稀少，从空中可以看到沙洲上的一小片树林，以及树林中高高的灯塔，还能看到庞大的珊瑚礁盘，礁盘的海水都不深，这些都是南中国海岛礁、沙洲的共同特点。

　　宣德群岛是属于西沙群岛的一组岛群，包括永兴岛、七连屿、宣德环礁、银砾滩等，位于最西端的西沙洲岛是七连屿的组成部分，座落在庞大的珊瑚礁盘上。西沙洲岛的沙滩是绿海龟上岸产蛋的好地方。

　　The West Sand Islands are piled with coral sands but sparsely covered with trees. Looking from the air, you can see a small stretch of woods and the tall beacon towering among the woods. Besides, you can also see huge coral reefs. These reefs lie in the shallow parts of the ocean. All these are the common features of reefs in the South China Sea and sandbanks.

　　The Amphitrite Group is a group of islands belonging to Paracel Islands, including the Yongxing Island (Woody Island), the Seven-sisters Islet, the Xuandehuan reef and the Iltis Bank. The West Sand Islands, which is situated on a huge reef in the western-most part of the Amphitrite Group, is part of Seven-sister Islet. The sandy beach of the West Sand Islands is a good place for green turtles (*Chelonia mydas*) to lay eggs.

海蚀洞•山东烟台
The Sea Chasm • Yantai, Shandong

  海岸边岩石较脆弱的部分经不住海浪的反复冲击会崩塌、破碎，并在岩壁上形成凹陷，进而形成洞穴。这些洞穴在海浪成千上万年地不断冲击下会越来越深、越来越大。画面中的海蚀洞就是这样形成的，其形状很有特色。

  海蚀洞与陆上的溶洞是有根本区别的。溶洞是水溶解石灰岩形成的，但海蚀洞是海浪机械侵蚀的结果。

  As a result of years of constant erosion by waves, certain fragile parts of rocks near the shore would crumbleand cave in to form grottos in the cliffs. These grottos will become larger under the impact from waves hundreds of thousands of years. The sea chasm displayed in the photo is shaped just through this process and looks extremely unusual.

  Sea chasms are totally different from the caves formed in inland places. Caves are often the result of limestone dissolved in water, but sea chasms are more often the result of physical erosions caused by waves.

蕈状石·台湾柳野
The Mushroom-like Stone • The Yehliu Geopark, Taiwan

　　在台湾的柳野海洋地质公园里能够欣赏到多种海蚀景观，其中以画面中的这类蕈状石最为有名。

　　岩石海岸在海浪长年累月作用下除了会形成海蚀洞外，还会造就出其他一些奇异景观，如海蚀沟、蜂窝石、烛状石、豆腐石、蕈状石、壶穴、溶蚀盘等。

　　You can see various marine abrasion landscapes in the Yehliu Geopark of Taiwan Province, the most famous one among which is the Mushroom-like Stone shown in the photo.

　　Besides sea chasms, rocky shorelines subject to long term wave erosion are likely to present viewers with some other special landscapes, such as, ocean-erosion ditch, alveolar stone, candle-like stone, tofu-like stone, mushroom-like stone, pot-hole and melting erosion panel and many others.

5 湿地与人类
Wetlands and People

　　由于地球水资源的全球分布在地理上和时空上的严重不均，干旱地区和湿润地区之间，河流上下游之间，山上和山下之间，北坡和南坡之间，旱季和雨季之间，这流域与那流域之间，等等，都存在严重的不平衡。因此，人类除了逐水草而居，靠水来人畜饮用、种植庄稼和发展生产以外，还根据社会经济发展的需要，修建了很多人造河流湿地，如水渠、运河等，修建了很多人工湖泊湿地，如水塘、水库等。并用自然保护地来保护天然的湿地，用湿地自然公园为人类游憩提供人造或改造的湿地。总之，无论湿地是天然的还是人造的，人类的生活、生产时时刻刻都离不开湿地。

　　Due to the imbalanced distribution of water resources on the earth in terms of geography, time and space, notable variations exist in the supply of water resources in different areas, for instance between the arid and humid areas, the upper and lower reaches of rivers, the top and foot of mountains, the northern and southern slopes, dry and rainy seasons, or between different river basins. Therefore, in addition to living by macrophytes, relying on water for drinking, feeding livestock, planting crops and improving production, human beings have also built many artificial riverine wetlands and lake wetlands, such as water channels and canals, ponds and reservoirs, according to the needs of social and economic development. All in all, be they of natural or artificial origin, wetlands are an indispensable part of the globe on which human beings rely heavily for their daily lives and productive activities.

藏族村寨和它的青稞地•西藏定日
Tibetan villages and highland barley •
Tingri County, Tibet

　　喜马拉雅山北坡海拔近4000米的地方，也许是人类村寨生活海拔高度的上限。这里土地贫瘠粗劣，生活环境条件很差。人们依靠从雪山流下的冰凉的溪水供人畜饮用、种植青稞来维持生活。

　　在如此恶劣的自然条件下，只要有了水，就有了生命，有了这里人们生活的全部。即将收获的一片片金灿灿的庄稼地，给这个贫瘠的山沟带来了全部的生气和希望。

　　The 4,000-meter-high places on the northern slope of the Himalayas might be the upper limit in altitude for human settlements and villages to be found. There are barren farmlands and poor living conditions here. People rely on cold streams originated from snow mountains for drinking, feeding livestock, planting highland barley (*Hordeum vulgare* var. *coeleste*) to sustain their life.

　　Under suchharsh physical environment, as long as there is water, there will be life and everything people live here. The goldencrops that are about to ripen soon bring all hopes for this barren village.

河曲牛羊多 • 新疆巴音布鲁克
The Copious Land of Cattle and Sheep in the River Bend • Bayanbulak, Xinjiang

----

巴音布鲁克，蒙古语意为"富饶的泉水"，位于天山腹地的和静县境内，著名的巴音布鲁克天鹅国家级自然保护区就座落在这片大草原中。

这里海拔近3000米，四周雪岭冰峰连绵，有充沛的泉水、雪水汇入大大小小的河中、湖中，水草丰美，牛羊马各种草原牲畜规模都不小。

Bayanbulak, meaning "copious springs" in Mongolian language, is located in Hejing County of the hinterland of the Tianshan Mountains. The notable Bayanbulak National Nature Reserve for Swans is situated in this vast grassland.

It is nearly 3,000 meters above sea level, surrounded by snow mountains and icy peaks. There are abundant springs and snow water flowing into large and small rivers and lakes. Due to the existence of copious water and abundant grasses, the breeding scales of various livestock in grasslands, such as cattle, sheep and horses, are large.

水天云光•浙江南麂列岛
The Water, Cloud and the Sky • the Nanji Archipelago, Zhejiang

----

　　在近海的南麂列岛山岙浅海，海水人工养殖非常发达，主要产品有虾蟹类、贝类及藻类等。一排排养殖带和一行行网箱在高天云光的映衬下，显得生机勃勃，一派美丽的滨海风光。

　　南麂列岛国家级自然保护区是中国首批5个海洋类型的自然保护区之一，也是中国唯一的国家级贝藻类海洋自然保护区，被誉为"贝藻王国"，被联合国教科文组织列为世界生物圈保护区网络。

The captive breeding industry on the shallow sea of the Nanji Archipelago is highly developed, yielding primarily such high-value products like shrimps and crabs (prawn-crab), shellfish and algae. Set against the beautiful clouds and clear sky above, the row upon row of aquaculture zones and cages appear to be steeped in the vibrancy of life. What a beautiful offshore view it is!

Nanji Archipelago National Nature Reserve is one of the first 5 ocean-themed Natural Reserves, and the only national shellfish and algae ocean Nature Reserve in China. It is known as "the kingdom of shellfish and algae" and has been listed among the UNESCO's (United Nations Educational, Scientific and Cultural Organization) World Network of Biosphere Reserve.

河畔鱼米乡·云南芒市
The Land of Fish and Rice ·
Mangshi, Yunnan

这里是素有"遮放谷子芒市米"盛誉的地方，人们理想中的"山水林田湖草"的美丽景观在这里集中呈现了出来。

芒市河从芒市和遮放两个坝子（即小盆地）中间缓缓流过，四周森林密布的山地储存的水不断汇入溪中、河中，大片的水稻田望不到边，傣族寨子在这片绿油油的大地上星罗棋布，人与自然和谐共处，暖意融融。

Reputed for the local-grown millet and rice indigenous to Zhefang that falls within its administrative region, Mangshi is a typical showcase for the ideal and idyllic home of human beings characterized by the perfect blending of "mountains, rivers, forests, farmlands, lakes and grasslands".

The Mangshi River flows leisurely between the two basins in Mangshi and Zhefang County. The water reserved in mountains around forest continuous flows into streams and rivers. We can't see the edge of large rice farmlands. Villages of the Dai ethnicity groups are scattered across these green areas. Humanity and nature live in harmony here.

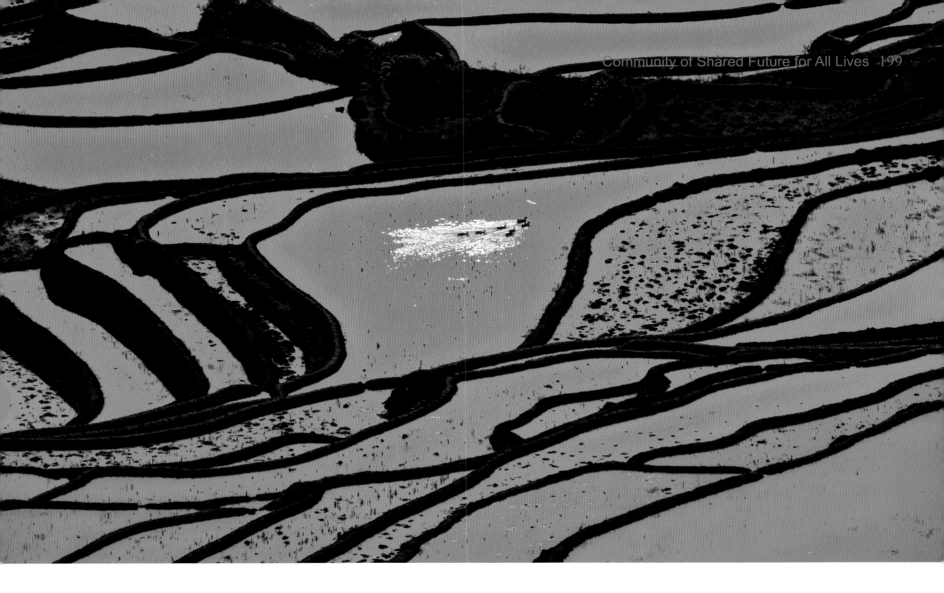

春田水暖鸭先知•云南哀牢山
The First Sign that Heralds the Arrival of Spring • Ailao Mountain, Yunnan

----

这里山高谷深、沟壑纵横、交通不便，但却蕴藏着哈尼人民在有限的山水之间创造的人间奇迹。

在村寨的上方，有着茂密的森林，提供着人们需要的水、用材、薪炭，下面是人们定居的村寨，村寨下方是层层相叠的千百级梯田，梯田里水稻、鱼、鸭共生。这一结构被盛赞为"林-水-寨-田-稻鱼鸭"五度同构的自然和人工的复合生态系统。哈尼梯田是人与自然高度协调的、师法自然的典范，属于人类社会共享的世界文化遗产。

In spite of the high mountains, deep valleys and ravines that make the Ailao Mountain extremely difficult to access through modern transportation facilities, this place is the home of a miracle created by the Hani ethnic community in defiance of the challenging natural settings for human survival.

High up in the mountains are lush forests that provide water, material, charcoals; down at the foot of mountains are the villages in which local people have their homes, tucked gently among overlapping terrace fields where rice, fish and duck coexist. This structure is known as the natural and artificial compound ecosystem in which the 5 composing elements—"forest, water, village, field and rice-fish-duck"—exist in perfect harmony. Hani's terrace field is a paradigm for making the best use of both nature and man-made environment for human wellbeing. It is a World Cultural Heritage shared by all humanity.

荒漠绿洲·甘肃敦煌
Oasis in the Desert • Dunhuang, Gansu

这在干旱的戈壁滩上，人工修建的一道道水渠有序地展开。沿着这些水渠，荒漠上出现了一片片绿洲，并依照水的力所能及，将树木、村庄、农田等不同绿色，在黄色的荒漠上有层次地铺展开来。

在干旱的大西北，有水就有绿色，有水就有生命，水和湿地比什么都重要。

Rows of man-made aquaducts extend out in orderly manners across the dry Gobi Desert. There are many oases along these water aquaducts, dotting around which are variously-colored trees, villages and farmlands that have been arranged across the yellow deserts.

In the arid northwest China, where there is water, there is greenness and life. Therefore, water and wetland are important things than everything.

江南古镇·安徽黟县宏村
Ancient Town in Jiangnan • Hong Village in Yi County, Anhui

宏村有近千年的村史，整个村子依山伴水，有先建水系后建村的前瞻，才使它有了水一样的灵性，被誉为"画中的村庄"和"中国最美丽的乡村"。

江南地区降雨充沛、河湖密布、水巷纵横，分布在这里的古镇是汉族水乡风貌最具代表性的村落，以其博大深邃的历史文化底蕴、清丽婉约的水乡古镇风貌、纯净古朴的民俗风情而闻名。而宏村更以湖光山色与层楼叠院和谐共处、自然景观与人文内涵交相辉映而驰名中外。

The history of Hongcun Village tucked among lush mountains and green water can be traced to thousands years ago. Thanks to the far-sighted vision of the earliest residents of the village who were bold enough to plan the water system prior to setting up the village, it was endowed with a unique ambience characteristic of water towns in Southern China. It is known as "village in picture" and "the most beautiful village of China".

Areas to the south of the Yangtze River tend to have abundant rain, rivers, lakes and water lanes. The ancient towns here are the most representative ones across the vast areas populated by the Han ethnic community. They are famous for their broad and deep-rooted history and cultures, the grace landscapes of ancient town and pure and simply folklore. In addition, Hongcun Village is world-famous for its co-existence between landscapes and villages, and the merge of natural scenery and culture.

密水浓云 • 北京密云水库
Dense Water and Clouds • Miyun Reservoir, Beijing

云彩密布的北京市密云区是北京最重要的饮用水源地和生态涵养发展区。在此修建的密云水库控制着潮河、白河共1.6万平方千米的流域面积，是京津唐地区第一大人工湿地，画面中间山顶上为长城烽火台。

南水北调工程的水进京后，担当了北京城区供水的主力军，减轻了密云水库的压力，水库蓄水量已经达到27亿立方米，水库水面达到188平方千米，创21世纪以来新高，同时也为解决华北地下水大漏斗问题作出了重大贡献。

Miyun District is the most important drinking water source area and ecology-conservation development zone of Beijing. Miyun Reservoir located here covers a total area of 16,000 square kilometers, over which the Chaohe River and the Bai River stretch. It is the largest artificial wetland in Beijing-Tianjin-Tangshan region. In the middle of this picture, we can see the Beacon Tower of the Great Wall on the top of mountains.

The South-to-North Water Diversion project, which has now been put under operation and become the primary source of water supply for urban Beijing, has more or less relieved the pressure that Miyun Reservoir used to be subjected to. Its water storage capacity has already reached 2.7 billion cubic meters and areas have reached 188 square kilometers, a new high since the 21st. At the same time, it has also made significant contributions to the cone of groundwater depression in North China.

鸟瞰都江堰・四川成都
A Bird-eye View at the Dujiangyan •
Chengdu, Sichuan

　　都江堰水利枢纽是两千多年前建成的水利工程，迄今一直在为人类造福。它是人类历史上伟大的水利工程之一，被列入中国最早的世界文化遗产和世界灌溉工程遗产。

　　目前，该灌区受益面积近百万公顷，涉及40多个县，使成都平原成为了旱涝保收的"天府之国"而富甲一方。这是中国人师法自然的一个相当成功的典型，功绩和影响彪炳千秋。

The irrigation system in the Dujiangyan is a water project built two thousand years ago and is stil benefiting the local community up to now. As one of the great water projects in world history, it was the earliest Chinese project listed into the World Cultural Heritage and the World Heritage Irrigation Structure.

Nowadays, the areas benefiting from this giant irrigation facility system have reached nearly one million hectares, including more than 40 counties. In addition, it has made the Chengdu Plain "land of abundance" with guaranteed harvests under drought and flood, and then become the richest region of Sichuan Province. It is the excellent archetype for following the laws of nature in China and its contribution and influence will always be keenly felt throughout history.

河套引水 • 内蒙古巴彦淖尔
The Water Diversion Project of the Hetao Area • Bayan Nur, Inner Mongolia

----

黄河河套灌区建有的以三盛公黄河水利枢纽工程为主体的引黄灌溉系统，灌溉面积达57.4万公顷。图为三盛公引黄灌溉系统的龙头工程，黄河从远方奔流而来，河水被水利枢纽一分为二，左边为黄河主河道，右面是人工开挖的引水渠，把黄河的部分水引入河套平原灌溉系统。

万里黄河进入蒙古高原后由北转向东，走出了黄河"几"字湾的第一个折弯，在折弯处产生了一个大大的河套。河套灌区水草丰美，物产丰富，是内蒙古难得的粮食基地，故有民谚"黄河百害，唯富一套"之说。

The Yellow River-diversion irrigation system in the Yellow River Hetao Irrigation Area, with the Sanshenggong-Yellow River irrigation system as its core, now covers an area of up to 574,000 hectares. This picture shows the most important project of this system. We can see the Yellow River running from far. The water is divided into two parts by the system. On the left is the main channel of the Yellow River; on the right is artificial water-diversion channel, diverting water from the Yellow River to the Hetao Plain's irrigation system.

After entering the Mongolian Plateau, the Yellow River turned from north to east, leaving behind it a huge loop that is shaped like the Chinese character "几". Due to the copious water, abundant grass, and sources, the Hetao Irrigation Area is a grain base in Inner Mongolia. Therefore, there is a folk saying, "The Yellow River has been flooded since ancient times, with the Hetao Area being the only place to reap its benefits."

夜临通州新城·北京通州
Nightfallat Tongzhou New Town ·
Tongzhou District, Beijing

这是正在建设的北京城市副中心，位于举世闻名京杭大运河之首的通州。夜幕降临，华灯初上，矗立在运河边古老的运粮楼在见证通州的新生。

京杭大运河始建于春秋，是世界上里程最长、工程最大的古代运河，也是中国人利用湿地、改造湿地，师法自然为人类造福的伟大工程。大运河贯通海河、黄河、淮河、长江、钱塘江五大水系，全长约1797千米。京杭大运河对中国南北地区之间的经济、文化发展与交流，特别是对沿线地区工农业经济的发展发挥了巨大的作用，被列入了世界文化遗产。

Tongzhou District, the newly established sub-center of Beijing this is still undergoing construction, is located in Tongzhou, the northern starting point of the world-famous Beijing-Hangzhou Grand Canal. When night falls and streets are bathed in pool of bright lights, the time-honored Daguang Tower, is witnessing the new stage of life of Tongzhou.

The Beijing-Hangzhou Grand Canal, as the longest and largest ancient canal in the world, was built during the Spring and Autumn Period. It is also a great project for Chinese people to make use of wetlands through taking advantage of their natural conditions for the benefit of mankind. The Canal, 1,797 kilometers in length, runs through five major water systems, including the Hai River, the Yellow River, the Huai River, the Yangtze River and the Qiantang River. The Canal plays an important role in the development and exchange of economy and culture, especially in the economic development of industrial and agricultural business along the regions of these rivers. In addition, it is also listed among the World Cultural Heritage.

中轴线上·北京朝阳区
The Central Axis • Chaoyang District, Beijing

北京奥林匹克森林公园位于北京的中轴线上，是为了承办2008年北京奥运会而人工修建的，挖湖垒山形成了著名的龙形水系和最高处的仰山。经过十多年的维护和演变，人工湿地已经变成了生物多样性丰富的半天然湿地，生长有丰富的湿地植物，常年有白鹭、野鸭、鹈鹕等鸟类出没。每年迁徙季节，灰雁、大天鹅等大型鸟类都会到这里觅食、歇息，因而北京奥林匹克森林公园被人誉为京北的"瓦尔登湖"。

到了2022年，北京又成了冬奥会的举办地，使北京成为世界历史上唯一的"双奥城"。

The Olympic Forest Park, located on the central axis of Beijing, was built as part of the city's bidding efforts to host the Beijing Olympiad 2008. The famous dragon-shaped water system and the highest Yangshan Hill in the Park are formed by digging the lake to pile up the hill. After 10 years of careful maintenance and natural evolution, the artificial wetland has become the semi-natural wetlands with abundant biodiversity. A rich variety of wetland plants and birds, such as egret, mallard, grebes (Podicedidae), are present all year round. During the migratory seasons every year, some large-sized birds such as greylag goose, whooper swan and other species will come here to for foraging and rest. Therefore, this Park is known as "Walden Pond" in the northern part of urban Beijing.

With the successful conclusion of 2022 Winter Olympics, Beijing has now become the only city among its world counterparts that has had the honor to host both the summer and winter Olympic Games—the City of Bi-Olympiads, as is proudly known among the local residents.

中国湿地保护与修复做出的主要成绩
Major Achievements that China Has Scored in Its Wetland Conservation
and Restoration Efforts

中国湿地保护与修复面临的挑战
Challenges that China Faces in Wetland Conservation and Restoration

中国湿地"十四五"展望
Prospects for the 14th Five-year-plan Period

生态思考

ECO-THOUGHTS

湿地是地球重要的自然资源和独特的生态系统，具有调洪蓄水、降解污染、净化水源、储碳固碳、调节气候以及涵养水源等生态功能，可以产出丰富的物质产品和生态产品，被誉为"地球之肾"，为经济、社会可持续发展提供着重要支撑。湿地还是众多野生动植物的栖息之地，是十分富集的"物种基因库"。贯彻实施《中华人民共和国湿地保护法》，更好地保护湿地生态系统和水资源，是维护国家生态安全和生物多样性的战略举措，是推动绿色发展和建设人与自然和谐共生现代化国家的迫切需要。

中国湿地保护与修复做出的主要成绩

湿地立法 规范行为

于2022年6月1日起实施的《中华人民共和国湿地保护法》是中国首部专门保护湿地的法律。该法立足湿地生态系统的整体性保护修复，确立了湿地保护管理顶层设计的"四梁八柱"。该法确立了"保护优先、严格管理、系统治理、科学修复、合理利用"的原则，建立了覆盖较全面、体系较协调、功能较完备的湿地保护法律制度，是推进新时代湿地保护高质量发展的重要保障，必将引领我国湿地保护工作全面进入法治化轨道。

湿地分类 夯实基础

按第三次全国土地调查汇总数据成果（2020年），中国土地分类的一级地类——湿地的面积为35204万亩（1亩=1/15公顷，以下同）。其中7个二级地类分别为：沼泽草地16716.22万亩、内陆滩涂8829.16万亩、森林沼泽3311.75万亩、沼泽地2905.15万亩、沿海滩涂2268.50万亩、灌丛沼泽1132.62万亩、红树林40.60万亩（不包括港澳台）。

按《湿地公约》口径，应再加上河流水面13211.75万亩、湖泊水面12697.16万亩、水库水面5052.55万亩、坑塘水面9627.86万亩、沟渠5276.27万亩，盐田933.54万亩，以及浅海水域5000万亩。

中国湿地总面积合计为87003.13万亩。

湿地保护 巩固成效

根据政府公布的数据：到2022年，中国政府指定了64处国际重要湿地、29处国家重要湿地，建立了602处湿地自然保护区、1600余处湿地公园。截至2021年年底，年度监测的63处国际重要湿地生态状况总体保持稳定，湿地总面积同口径相比有所增长，大部分湿地补水量稳中有升，总体水质呈向好趋势，生物多样性丰富度有所提高，湿地保护率提高到50%以上，新增和修复湿地面积80余万公顷。

至2022年9月止，有2批13座中国城市获得《湿地公约》批准的全球首批"国际湿地城市"称号。国家湿地公园已遍布全国31个省（自治区、直辖市），总数达899处，有效保护了240万公顷湿地，带动区域经济增长500多亿元。目前，约90%的国家湿地公园向公众免费开放，成为人民群众共享的绿色空间和"绿水青山就是金山银山"理念的生动实践。

湿地履约 勇担责任

我国首次举办的《湿地公约》第十四届缔约方大会（2022年11月），主题为"珍爱湿地，人与自然和谐共生"。这次会议对于加强中国湿地的保护和修复工作，实施好《中华人民共和国湿地保护法》，参与和引领国际湿地保护和修复工作，彰显中国推进构建人类命运共同体的决心，树立负责任的湿地大国形象，将具有重要的推动作用。

中国湿地保护与修复面临的挑战

尽管我国的湿地保护已经取得了很多成绩，但是在新时期，湿地保护和修复还面临着诸多挑战，主要有：

（1）社会对湿地生态功能和价值依然认识不足；

（2）破坏和过度利用湿地的现象依然存在；

（3）对"水是湿地的灵魂"的认识还没有提高到应有的高度；

（4）湿地修复中存在重人工干预、轻自然恢复的倾向；

（5）湿地保护与修复的法律法规需要配套落实执行到位；

（6）湿地保护与修复关键技术及典型示范不够全面；

（7）科研、监测及队伍建设尚不适应形势要求。

中国湿地"十四五"展望

《"十四五"林业草原保护发展规划纲要》中提出，中国湿地"十四五"的目标是，以第三次全国土地调查汇总数据成果（2020年）为基础，科学确定湿地管控目标，科学划定纳入生态保护红线的湿地范围，优化湿地保护体系空间布局，形成覆盖面广、连通性强、分级管理的湿地保护体系。

强化江河源头、上中游湿地和泥炭地整体保护，加强江河下游及河口湿地保护，在已建设湿地自然保护地的基础上，新建一批湿地类型的国家公园。

采取"近自然"的方案和措施，增强湿地生态系统自然修复能力，重点开展生态功能严重退化的湿地生态修复和综合治理。加强重大战略区域湿地保护和修复，实施红树林保护修复专项行动。

加强湿地管理，建立完善的湿地保护部门协作机制。完善湿地管理体系，建立健全湿地分级分类管理制度，发布重要湿地名录，制定分区管控措施。开展国际重要湿地、国家重要湿地的生态状况、治理成效等专题监测。

力争到"十四五"末，中国湿地保护率提高到55%，恢复湿地100万亩，营造红树林13.57万亩，修复红树林14.62万亩的目标。

Wetlands are a sort of unique ecosystem as well as a critical natural resourceof the Earth, playing an active role in flood regulation, pollution alleviation, water purification, carbon sequestration, climate change mitigation and water resources conservation. In recognition of the abundant material and ecological products derived from them, wetlands are sometimes compared to the "kidney of the Earth" that underpins the sustainable development of economy and society. Moreover, wetlands also provide safe shelters to a rich variety of wildlife flora and fauna species and therefore can be regarded as the "gene-banks of species". Earnest implementation of *the Law of Wetland Conservation of the People's Republic of China* is not just a strategic measure that safeguards the wetland ecosystems, the water resources and hence maintaining the country's ecological security and biodiversity, but also an urgent need of the country to promote green development and bring into reality a modern China characterized by the harmonious co-existence of people and nature.

Major Achievements that China Has Scored in Its Wetland Conservation and Restoration Efforts

Strengthening Regulation through Legislation

The *Law of Wetland Conservation of the People's Republic of China*, which effected on June 1, 2022, marks the first law of the country that themes specifically on wetland conservation. Conceived of from an overriding philosophy that advocates taking a holistic approach in the conservation and restoration of wetland ecosystems, the Law sets out an effective framework for exercising top-down management of all initiatives in this field. Built on basis of the central principle that highlights "conservation priority, rigorous management, systematic control, science-based restoration and rational utilization", a sophisticated legislative mechanism is now in place for ensuring that comprehensive, well-coordinated and task-oriented conservation projects can be carried outto their best effects. The debut of the Law marks a major stride ahead in China's law-guided wetland conservation endeavor and constitutes a reliable guarantee for sustaining high-quality wetland ecosystems in the new era.

Laying a Solid Foundation through Science-based Wetland Classification

Data generated from the third national land survey (2020) indicates that the total size of wetlands, which belongs to Category 1 land according to China's national land classification system, amounts up to 352.04 million *mu* (1 *mu*=1/15 hectares, the same below). Of the 7 Category 2 wetland types, their respective sizes are as follows: marsh grasslands, 167.1622 million *mu*; inland tidal flats, 88.2916 million *mu*; forest marshes, 33.1175 million *mu*; marsh lands, 29.0515 million *mu*; coastal tidal flats, 22.685 million *mu*; shrub marshes, 11.3262 million *mu*; and mangrove, 0.406 million *mu* (excluding those of Hong Kong, Macao and Taiwan).

If the system of the *Convention on Wetlands* is adopted as the standard, the following additional figures will need to be added to the list: rivers, 132.1175 million *mu*; lakes, 126.9716 million *mu*; reservoirs, 50.5255 million *mu*; ponds, 96.2786 million *mu*; ditches, 52.7627 million *mu*; salt pans, 9.3354 million *mu*; and shallow seas, 0.50 million *mu*. Putting all these figures together, we will see that the total area of wetlands in China stands roughly at 870.0313 million *mu* in size.

Consolidating Achievements Already Made in Previous Conservation Endeavors

According to the official data, as of 2022, the Chinese government had, in addition to the 602 wetland nature reserves and more than 1600 wetland parks that had been set up, designated 64 wetlands that are of international importance and 29 other ones that are of national importance. By the end of 2021, of the 63 wetlands that are of international importanceand closely monitored on annual basis, the overall ecological wellbeing has basically remained stable and healthy, with their total size increased slightly compared with previous data. As the water supplement for most wetlands is on increase steadily and overall water quality is showing encouraging trends for improvement, the biodiversity of the wetland ecosystem has also notably improved. In summary, over 50% of existing wetlands has been put under effectiveprotection, with another 800,000 hectares newly established or recovered.

As of September 2022, 13 Chinese cities have been consecutively honored by the *Convention on Wetlands* the "International Wetland Cities". The total number of National Wetland Parks, which are extensively distributed acrossthe country's 31 provinces (autonomous regions and municipalities directly under the Central Government), has reached 899, putting 2.4 million hectares of wetlands under effective protection and generating a 50-billion-yuan-worthy growth in regional economy. At present, about 90% of the National Wetland Parks are open to the public free of charge, which, besides making them green spacesreadily accessible to the people, also testifies powerfully to the valid of the popular saying that "green mountains and lucid waters are comparable to mountains of gold and silver".

Greater Responsibilities in Fulfilling Its Obligations to the *Convention on Wetlands*

China will for the first time host the upcoming *Convention on Wetlands* COP 14 (Nov., 2022), the theme of which is "Wetlands Action for People and Nature". In addition to boosting China's confidence in doing a good job in implementing the *Law of Wetland Conservation of the People's Republic of China* and in strengthening its endeavors in this field, the conference will also play a conducive role for China to assume greater responsibilities and take initiative in the efforts of the international community in wetland conservation and restoration. It will demonstrate to the world China's strong commitment to building a global community of shared future for human beings.

Challenges that China Faces in Wetland Conservation and Restoration

In spite of the encouraging achievements that China has made, we are still confronted with many grave challenges, as listed below:

(1) less than satisfactory appreciation of the eco-functions and values of wetlands on the part of the public;

(2) lingering abusive and even detrimental utilization of wetland resources;

(3) insufficient awareness about the meaning of the doctrine that "water is the soul of wetlands";

(4) tendency to valueman-initiated intervention over natural recovery in wetland restoration projects;

(5) laws and regulations on wetland conservation and restoration yet to be further streamlined or optimized;

(6) key technologies and pilot programs yet to be further developed and extensively promoted;

(7) infrastructure for scientific researches, monitoring and capacity building yet to be upgraded to better meet the demands of the new era.

Prospects for the 14th Five-year-plan Period

The *Framework of Forestry and Grassland Protection and Development for the 14th Five-year-plan Period* identifies the objectives of the country's wetland conservation efforts during the upcoming 5 years as follows: through drawing on the results yielded from the third national land survey (2020), determining science-backed goals for wetland management and control; delineating wetlands that should go within the ecological protection redlines set according to rigorous and proven standards; optimizing the spatial distribution of the wetland conservation system; putting in place an extensive-covering, well-coordinated and reasonably-tiered management system for wetland conservation.

Priority will be given to the comprehensive conservation of wetlands and peatlands situated at the source, the upper and middle reaches of the country's major rivers, while strengthened efforts will be made for the conservation of wetlands located at their lower reaches as well as the estuaries. Some new national parks featuring wetlands will be built to further supplement the existing wetland nature reserves.

"Near nature" approaches will be adopted to foster the self-recovery capacity of wetland ecosystems, highlighting in particular the comprehensive management and restoration of wetlands suffering from serious degradations in terms of their ecological functions. In addition to the conservation and restoration of wetlands that are of strategic importance, purposed-designed special programs will alsobe put under operation for the protection and restoration of mangroves.

Besides putting in place a sophisticated mechanism for streamlining cross-section cooperation in wetland conservation initiatives, efforts will be made to set up a comprehensive wetland management system that will be responsible for effecting tier-specific, category-specific, zone-specific supervisory measures as well as for releasing lists of critical wetlands. Purpose-designed special programs will be launched to keep track of the well-beings of wetlands that are of international and national importance, as well as of progresses made in their conservation.

The overall goals are that, by the end of 14th Five-year-plan Period, 55% of the country's wetlands will be placed under sound protection, with an additional 1 million *mu* of wetlands duly recovered, 135,700 *mu* of mangroves newly established, and another 146,200 *mu* of degraded mangroves that are currently in existence restored to healthy status.

后记 EPILOGUE

为迎接《湿地公约》第十四届缔约方大会（COP14）在中国武汉的召开，中国林业出版社向我约稿一本湿地主题的生态摄影作品，展示中国湿地之美，科普湿地方面的知识。8年前，我曾出版过《多样性的中国湿地》，这次即将出版的这本生态摄影集，既要与另外两本成一个系列，又要与我之前出版的有区别，难度着实不小。幸好我有几十年工作经验与摄影爱好相结合的大量积累做基础，再加上平时生活中习惯观察与思考，经过努力，书稿终于如期交付了。

生态摄影是集"思想性、科学性、艺术性"为一体的艺术形式，是当今生态文明社会建设的产物，生态摄影要求用图片及文字来讲述精彩的生态故事，为科学普及和自然生态教育提供最有力和大众最喜闻乐见的平台。在这本书中，我依然秉承这个生态摄影理念，精心挑选图片，斟字酌句地撰写图片说明，但是由于本人知识和能力有限，书中论述难免有错误和不当之处，图片选用方面又总会在光影构图好和说明问题之间犹豫而留下遗憾。总之，我诚挚地希望读者们多多提出批评和宝贵建议。

通过写这本书，我深深地体会到平时的思考和资料图片的积累是多么的重要。因此，我要感谢几十年给我支持和帮助的国家及地方（自治区、直辖市）自然保护管理部门及基层工作的同仁们。此书能够在短时间里圆满出版发行，特别要感谢中国林业出版社成吉书记的全力支持，国家林业和草原局湿地司吴志明司长、陈良副司长的及时指导，以及肖静、袁丽莉、崔林、田红等编辑的大力支持，还要感谢我的家人对我这个退而不休的老人的倾心理解和支持。在此，我一并表示衷心地感谢！

陈建伟

2022年6月

Delighted as I was to receive the invitation of China Forestry Publishing House to write a new eco-album that, in addition to celebrating on the beauty of wetlands in China and playing a role in disseminating knowledge concerning wetlands, will be used as a tribute to the upcoming *Convention on Wetlands* COP 14 in Wuhan, it is by no means an easy job for me. The reason is that I have in 2014 published another eco-album, the *Diverse Wetlands of China*, that features on the same topic. This newly planned eco-album is supposed to be a follow-up to the previous work yet notably different from it, which is indeed a challenging task. Fortunately, thanks to the rich experience I have gained over the past several decades and my bounty collection of photos captured while working in the field, as well as to the helpful habits I have developed over the years in keeping an observant eye and a mindful thought on daily life, I somehow managed to deliver the manuscript to the publishing house in time.

Eco-photographing, an art that blends ideas, science and artistic tastes into an integrated whole, is a product of the present age when ecological progresses are regarded a symbol of a civilized society. Its primary goal is to take advantage of vivid pictures and brief, thought-provoking languages to tell stories that celebrate on ecological attainments, hence creating an effective yet popular platform for spreading knowledge and boosting nature education. I have consistently adhered to the eco-photographing ideas that I believe to be true by taking great pains in the selection of each picture used in the book and each word and phrase chosen for the captions. Nevertheless, due to the limits of my own knowledge and pressure of time, there inevitably might be some mistakes or inadequacies in the book. The painful tradeoffs I have been frequently faced with in balancing the visual merits of a photo and its ecological values might inevitably result in regrets. But anyway, it's sincerely hoped that my readers will not hesitate to feed me back with their constructive comments and valuable suggestions.

The writing of this book once again reminds me what a critical role that rigorous thinking and timely documentation in daily life play in the making of a ready writer. For this reason, I would like to extend my most heartfelt thanks to all my colleagues working in both the central and local nature conservation management agencies who have constantly afforded me their utmost supports over the past decades. My special thanks go to Mr. Cheng Ji, the CPC secretary of China Forestry Publishing House; to Director Wu Zhiming and Deputy Director Chen Liang from the Department of Wetlands in China's National Forestry and Grassland Administration; as well as to editors Xiao Jing, Yuan Lili, Cui Lin and Tian Hong. Hadn't it been for the devoted work and whole-hearted supports of them, it couldn't have been possible for me to complete this book and have it published within such a tight schedule. Last but not least, my most cordial thanks go to my family members, whose unreserved supports and understanding care have helped me to complete this book in my retirement. A most sincere thank-you to you all!

Chen Jianwei

2022/06